THE
WORMWOOD
PROPHECY

THE
WORMWOOD
PROPHECY

THOMAS HORN

CHARISMA
HOUSE

Most Charisma House Book Group products are available at special quantity discounts for bulk purchase for sales promotions, premiums, fund-raising, and educational needs. For details, call us at (407) 333-0600 or visit our website at www.charismahouse.com.

THE WORMWOOD PROPHECY by Thomas Horn
Published by Charisma House
Charisma Media/Charisma House Book Group
600 Rinehart Road, Lake Mary, Florida 32746

This book or parts thereof may not be reproduced in any form, stored in a retrieval system, or transmitted in any form by any means—electronic, mechanical, photocopy, recording, or otherwise—without prior written permission of the publisher, except as provided by United States of America copyright law.

All Scripture quotations are taken from the King James Version of the Bible.

Copyright © 2019 by Thomas Horn
All rights reserved

Visit the author's website at www.skywatchtv.com.

Library of Congress Cataloging-in-Publication Data:
An application to register this book for cataloging has been submitted to the Library of Congress.
International Standard Book Number: 978-1-62999-755-1
E-book ISBN: 978-1-62999-756-8

While the author has made every effort to provide accurate internet addresses at the time of publication, neither the publisher nor the author assumes any responsibility for errors or for changes that occur after publication. Further, the publisher does not have any control over and does not assume any responsibility for author or third-party websites or their content.

19 20 21 22 23 — 987654321
Printed in the United States of America

CONTENTS

AUTHOR'S NOTE

MANY PEOPLE IN the media and pop culture use the terms *asteroid*, *meteoroid*, *meteor*, and *meteorite* interchangeably, but NASA has distinguished among the four, defining them each as slightly different items. An asteroid is a formation of rocky substances, often made of silicate, clay, and varying metals, most often nickel-iron.[1] Meteoroids are created when asteroids collide with each other, creating broken fragments of the previously larger bodies.[2] When these displaced remains enter Earth's atmosphere and vaporize, they create what many regard as shooting stars, more scientifically known as meteors.[3] When one of these bodies does not completely dissipate in our atmosphere and makes contact with the earth's surface, it is then called a meteorite.[4]

Comets, on the other hand, are composed of ice and dust, and as they travel near the sun, the ensuing heat vaporizes the ice and dust, creating a tail.[5]

Because, aside from technical, scientific works, these terms are used interchangeably (and to avoid redundant use of the same term), it is assumed that the reader can likewise apply these terms similarly using the context wherein they are applied within this book. Thus, within this work the terms *asteroid*, *meteor*, and *meteorite* will all refer to an orbiting mass of rock either in space or having already collided with Earth at some previous time.

ALTHOUGH THIS BOOK is nonfiction, I begin with the following narrative, then end this work by returning to this imaginary story line. The reason I have chosen this unusual methodology is to illustrate as vividly as possible what I believe to be at play behind the scenes, which will become obvious as the reader moves through the factual portion of the manuscript, starting with chapter 2.

—THOMAS HORN

Chapter 1

WELCOME TO WONDERLAND

P ETE McCUTCHEON FELT decidedly underdressed. At 3:33 that afternoon, two uniformed men called at his modest home near Marionville, Indiana, and hustled him into a black SUV, then into a helicopter, and finally to a sleek air force jet, which transported its befuddled passenger to a top-secret airport high in the Colorado mountains.

After passing through a series of checkpoints and metal detectors, McCutcheon followed his close-mouthed captors into an elevator and descended seven stories below ground level. A third man met them once they left the elevator's stainless steel interior, but this one actually smiled.

"Good evening, Major McCutcheon. Welcome to Cheyenne Mountain."

"I'm retired, Captain," Pete replied with a raised eyebrow. "Heart condition. It's been fixed, but I'm not the able-bodied man I used to be. I just write, play with my grandkids, and give the odd talk now and then."

"Yes sir, we're aware of that," the officer told him as they passed along a series of corridors. "I want to apologize for stealing you away from your family without notice. One of our team members explained it all to your wife and in-laws shortly after you left. We're meeting in here, sir," he finished as they approached a nondescript metal door flanked by a pair of bulky men in air force blue.

The taller of the two whispered, seemingly to no one, and then nodded his head. "They're ready for you, Captain Andrews."

"This way," the polite Andrews told Pete. "You're about to step through the looking glass, sir. Better buckle up."

***

Seventeen hours away, on the channel island of Jersey, fifteen-year-old Ben Brandeis stared at his computer screen. He held half a Marmite sandwich in one hand, a plastic tumbler of fizzy orange drink in the other. He really should have been sleeping, but the alarm on his search for extraterrestrial intelligence (SETI) program had awoken him half an hour earlier. Ben shoved a pair of wire-rimmed spectacles up on his nose, blinking to clear his myopic vision. The previous summer, his father had agreed to move the teen's room into the attic, allowing Ben privacy and a bit of quiet, so he felt certain his movements hadn't disturbed his hardworking dad's own rest. Bill Brandeis worked as a lorry driver for one of the big French shipping companies, and he'd be rising soon—heading out to fetch the first load of the day.

But why had the SETI alarm sounded? Ben ran a freckled hand through a thick thatch of unkempt auburn hair, his hazel eyes scanning row upon row of charts and numbers that made sense to very few of his friends. With an IQ of over 175, Ben had no trouble with coursework, but even his A-level instructors found the teen's intensity difficult to comprehend.

Deciding the numbers couldn't be right, Ben switched the monitor's input to a second, more powerful computer and opened an astronomy program he'd written the previous spring as part of an application to Oxford. Two weeks later he'd received a personal visit from Dr. Alan Holmes, the astrophysics chair of Merton College, part of Oxford, and the director of the sky-mapping program C-BASS. Holmes had offered the lad a place in his personal lab as well as free tuition and board. The word *genius* peppered the lean professor's conversation, and Bill Brandeis beamed with pride as he'd shaken the visitor's smooth hand, explaining how his late wife—Ben's mother—would have been pleased as punch to know how well their only child had turned out.

Ben had loved his mother dearly and as a tribute—following her death to cancer—had named his own near-Earth-object tracking program after her, NEO-Stella. Thinking of her, Benjamin smiled softly and clicked a wireless mouse to open the complicated program. A series of colorful icons appeared on the flat-screen monitor, giving links to six telescope arrays for viewing. Before meeting with Holmes, Ben had hacked into these installations surreptitiously, but now he could use a personal security code, which offered the boy genius access to any of the stations within the C-BASS alliance. The intricate code in the Stella program tapped into the massive array at Owens Valley Radio Observatory in California, but finding nothing unusual, Ben switched to the Hartebeesthoek Radio Astronomy

Observatory in South Africa, now focused on a newly discovered pulsar within the Trapezium Cluster of the Orion Nebula. When the SETI program on his older machine had alarmed at two o'clock that morning, Ben noticed it fixed upon the great hunter's constellation, and the brilliant teen had a hunch that whatever tripped the alert lay within Trapezium.

By 4:10 that morning, in mid-January, young Benjamin Brandeis would find out why—and it was a shock that would soon ripple throughout every world government.

"Sit, Major," a middle-aged woman in a gray suit said as McCutcheon entered the room. The air smelled like hand sanitizer mixed with heavy floral, and Pete noticed vented fans positioned near the ceiling of each wall. A faint, low-pitched hum caught his attention—as did four closed-circuit television cameras, mounted in each corner. The woman laughed. "Don't worry, Major McCutcheon, none of this is being recorded—not on the cameras, anyway. My secretary here, Lieutenant Evans, has an eidetic memory, and he's keeping track of everything we say. You'll understand in a moment just why none of this will be logged. Not yet."

Pete took one of the metal and vinyl chairs opposite Evans, a thin-lipped man with a boyish face and protruding ears. Pete had a hunch that the eidetic memory had more to do with a brain chip than any native ability to recall with clarity. "And you are?" he asked the woman.

She smiled again, the slow curve of her upper lip revealing a slight dimple in her left cheek. Despite the pleasantries, the woman had cold eyes, chillingly frank and logical. "Dr. Gale Stone," she answered. "I work with a team of scientists whose names you already know. Major, it's—"

"I don't use the military rank any longer, Dr. Stone. Just call me Pete—or Dr. McCutcheon, if you prefer. Most of my friends just call me Mac."

"Forgive me," she said without a blink. "Mac, then. I'm familiar with your work at Caltech, but more to the point, the recent book you wrote on biblical prophecy." She snapped her fingers, and Evans handed her a thick hardcover text in a colorful dust jacket. Dozens of paper strips marked pages. "This is hardly light reading," she said as she opened the book to one of the marked passages. "In this day and age I wonder how you managed to get anyone to publish it. Tell me, Mac, do you really think a star called Wormwood will one day strike the earth?"

"Is that why I'm here?" he asked in irritation. "Look, Dr. Stone, if you're looking to make fun of my beliefs, then you've caught the wrong man. I make no bones about my stance on eschatology— neither as a Bible scholar nor as a scientist."

"Yes, but it's that scientific background that most interests me," she interrupted. "Mac, I've not brought you halfway across the country to belittle your beliefs. Rather, I respect and appreciate them. You are uniquely qualified to provide counsel on a matter that threatens not only the United States but the entire world. Interested?"

The retired major took a moment to respond. The guard near the door continued to whisper into his internal communication system (which Mac later learned was a two-way radio mounted on one of the young man's back molars), and he nodded now and then as if receiving orders. The guard never smiled. Mac had an unsettling spidey sense that he'd not only passed through the looking glass but fallen down a deep hole. If so, then this woman would be either the White Queen or the Red Queen.

He silently prayed to keep his head.

"Dr. Stone, if you've read my dossier," he told her, "and it's obvious you have, then the answer is yes, providing you agree to tell me everything about this alleged threat. I don't like working in the dark. If I'm to offer counsel, then I require full disclosure. If you can't make that promise, then I'll waste no more of your time. As you know, my in-laws are visiting, and we had several post-Christmas activities planned, including taking in a show in Indianapolis."

"Very well." She pushed the book toward him, open to page 231. "Wormwood. That's what's coming. And it is probably going to strike our planet in less than six hundred days. If you want your new granddaughter to reach her seventh birthday, then we need you, Mac. I have two children myself, ten and thirteen. I'd like to see them married, watch them go to college, maybe have kids of their own, but if we can't find a way to stop this thing from crashing into us, then mankind can kiss its collective posterior goodbye."

The book was open to the chapter he'd written called "The Sword of Orion," where he discussed the possibility of the sudden emergence of a near-Earth object, or NEO, from within the Orion Nebula.

He sighed. "Why me? There are loads of other prophecy scholars with better reputations than mine. I only got into this five years ago, after I retired from the air force."

"But no one with your unique set of skills, Mac," she argued. "You have doctorates in astrophysics and mathematics, but you also finished a second bachelor's in genetics and nearly completed a third in computational theory. You are what we call a polymath—a brilliant mind with the ability to connect disparate disciplines to solve a problem. That's why the force worked so hard to keep you, and it's why we need you now. President Hernandez himself requested you, Mac. He read your book, and he's a believer in what you say."

Peter McCutcheon thought of his wife, Amelia, and their two

children, both married with small children—his precious grandchildren. He considered the conclusion of his most recent manuscript—as yet unpublished—where he wrote that fluctuations in the Orion Nebula might indicate the Wormwood of the Book of Revelation might actually arise without warning from that birthplace of the stars. A clawing sensation crept into his stomach and radiated along his spine. He glanced up at the blinded cameras. A sudden desire to escape overwhelmed his spirit, but a small voice whispered that he *had* to do this. He'd been designed to do this.

"You were set apart before you were born," it whispered, "and called by My grace."

He swallowed the fear and anxiety, wondering if she could see his hands shaking.

"Tell me what you want me to do."

On Jersey, Benjamin Brandeis furiously made notes in a spiral-bound notebook, coded so no one else might discern their meaning. His hands danced along the pages, forming a series of nonsensical letters and numbers based on a deeply encrypted cipher he'd designed when only eleven years old. Line upon line consumed the paper, interrupted now and then by complicated equations.

As he finished the calculations, Ben gasped in shock at what he saw.

"That can't be right," he said aloud, his heart thumping. "It's changed trajectory. How could it have done that?"

After tucking the notebook into his backpack, Ben thumbed a quick text to his dad. If he caught the ferry to Poole, then Ben could take a train to Oxford and be there by two that afternoon. He felt sure Dr. Holmes would be willing to meet him, but if not, he'd board another train for London. No matter what, he had to present these findings to Holmes in person. If he was right—and Ben felt certain he was—then the world had less than three years before something massive would hit it. A planet killer.

Life as he and his school friends knew it was about to change.

ONE WEEK LATER...

Retired major Peter McCutcheon extended a hand to the youth with the auburn hair. "Welcome to Wonderland, Ben," he told Brandeis.

The teen blinked. "Sir?"

Pete laughed. "It's what we call our team. Military code names seldom use anything that might give us away. We're anxious to hear your report. Is Dr. Holmes with you?"

"Downstairs, sir. He had a hiccup with a belt, apparently. The metal buckle triggered the alarm. Is the president really going to be here?"

"Later," Mac answered as he led the boy genius into the White House meeting room. "We're soundproof in here. The press corps has been told the president is congratulating you on the discovery of a new comet. Of course, you and I know what you really found."

"I'm not sure anyone understands just how dangerous this is, Dr. McCutcheon. It's changed trajectory twice now."

"Actually, three times," Mac countered. "We noticed an aberration to its path this morning at 5:13 (GMT). It's passing through an asteroid field, we think. These are striking it like obstructions in a pinball game, and it's ricocheting. Our people feel there's better than a 39 percent chance it will strike somewhere in the northern hemisphere."

The rectangular table that dominated the conference room had a large, circular screen mounted to the surface. Twenty chairs surrounded the table, six of them filled with men and women in dark suits and even darker faces. Four of the chairs held men in uniform, all with graying temples, if not entire heads of silver. A youthful-looking fellow in a lighter suit appeared to be making notes in a leather journal, and a woman with salt-and-pepper hair swept into a spiraling chignon walked to greet Pete and Ben.

"Hello, Ben," she said with a pleasant smile. "I'm Dr. Gale Stone. We're delighted you agreed to join our team."

Young Brandeis swallowed, the gesture causing his Adam's apple to bob along the track of a long throat. He extended his right hand in response and shook hers. "Thanks," he managed to reply. "I never imagined I'd be sitting inside the White House. I visited Buckingham Palace once, on a field trip for school."

"So I've read," she answered with surprising geniality. "Let's all sit, shall we? There's a lot to cover and very little time. We'll be serving lunch in here, so make yourself comfortable, Ben."

He took a seat beside the man with the leather journal, with McCutcheon on the other side. The older man offered encouragement. "I know you're feeling out of your depth, son," he whispered as an air force sergeant started to shut the door. "We're all friends here."

Outside, a man exchanged words with the clean-shaven guard, and a fortyish man with shocking red hair entered, instantly dominating the room.

"Sorry to be late," Holmes stated, a grin widening on his bright face. "I keep forgetting about this blasted belt of mine. The buckle's made of a rare metal alloy that always concerns these detectors. Ben, I see you've already met Pete McCutcheon. Nice to see you again, Mac. I say, did anyone order up tea?"

The Oxford professor took a chair to the right of Dr. Stone, and over the course of the next five minutes the table continued to fill with men and women of all disciplines. Once the team was complete, Dr. Stone nodded to the guardian of the door. "That will be all, Sergeant Williams. We'll signal when we're ready for refreshments."

The officer nodded and left, locking the door from the outside.

The circular screen that overhung the table flickered to life, and a series of images appeared, featuring the words "Wonderland Strategic Group" layered over the stylized graphic of a streaking object that intersected the letters between "Wonderland" and "Strategic" as if crashing. Ben Brandeis flinched when he saw the disturbingly cheerful logo, amazed that a designer had actually been paid to draw an emblem depicting the end of the world.

As if she read his mind, Dr. Stone offered an explanation. "As you will see, ladies and gentlemen, we now have an official look for the Wonderland program. Though it may seem lighthearted to provide a branding to our initiative, I assure you that the effort is one of misdirection. Should anyone ever discover any of our pages within the nested websites or on a forum populated by our paid provocateurs, they will find nothing but a series of misleading slides and documents. The world must not suspect what we are about to discuss. If it did, panic would seize the hearts of more than seven billion people, and that is a scenario we can never allow."

She paused a moment to open a thick packet of printed material that matched portfolios set before each of the members. "If you will turn to page 10, we'll begin with a short summary of our mission. A few of you haven't yet met our external liaisons. Major Peter David McCutcheon retired six years ago, but we've reactivated his commission and named him our official scientific adviser. Mac's academic prowess is legendary, but he's also an expert on the prophetic aspects of Wormwood that may help us unravel and even prevent what is happening to us."

A middle-aged man in a blue suit sitting near the end of the table glowered, his puffy eyes downturned, his hands steepled. "Bunch of nonsense!"

Dr. Stone's reply cut the man down to size. "Secretary Gilmore, I'm all too familiar with your opinion regarding my choice to include Major McCutcheon, but the president has overruled your objections—as you know very well. Now our other guest has made a favorable impression on all of us. Benjamin Algernon Brandeis is a high-level genius with an unparalleled mastery of astrophysics and an understanding of mathematics that rivals that of Einstein. Never in all my years with the Cheyenne Mountain Division have I met so gifted an individual. I suspect Major McCutcheon would describe him as one of God's ministers, chosen for such a time as this. Ben, welcome to Wonderland."

Brandeis felt a chill pass through his frame as every eye cast its gaze upon him. A few seemed receptive, even kind, but most held sparks of jealousy and doubt. He could tell the people thought him misplaced, a youthful upstart, an intruder, little more than a boy trying to play with grown men.

"Thank you, Dr. Stone," he answered, praying his voice wouldn't crack. "Dr. Holmes is the one who's made a lot of the breakthroughs on this."

The Oxford chair disagreed. "Ben's a humble sort o' chap," he told the group. "I'd never have come up with these calculations on my own. In fact it's Benjamin's computer program that C-BASS is using to keep track of Wormwood."

"Wormwood?" the teen asked. "Is that what you're calling it?"

McCutcheon answered, "It's how God described it. It means bitterness. May I?" he asked Dr. Stone.

She nodded. "Yes, of course."

Taking up a remote control, Mac continued as he stood to address the gathering. "Most of you know me from my previous stint with the Force. And for the past week many of you have toiled beside me as we've tried to figure out just what is happening to the NEO we're calling Wormwood. But for the few of you who are joining us for the first time—and I speak to Secretary Gilmore and Senator Malone, in particular—let me give you a quick summation. In

June of 2004 a near-Earth-object scanning program called NEON mapped the location of a mass in the Orion Nebula, designated as 2004 JU04. This object behaved normally until late 2015, when it suddenly brightened. Spectrometry performed at JPL indicated JU04 might be a red giant, despite its white appearance, and in response the European Space Agency (ESA) launched a probe to discern its composition. Early results came in around six years ago, revealing numbers that made no sense whatsoever. In response a joint ESA and NASA team was formed to track it, and a second probe was launched to obtain new measurements. About that time, a cover story was released to all the usual outlets that a well-known tech tycoon was planning to send passengers to Mars and using launches from ESA and NASA to pave the way for that effort."

Nearly everyone in the room smiled or laughed softly, for all Washington insiders knew the "tycoon" to whom Mac referred. In fact that very man would have been at the table that day were he not in China working on an ancillary project: the final phases of a smart city, meant to house the elite from the wrath of Wormwood.

McCutcheon continued. "As I said, the object was thought to be destroyed. All that changed three years ago, when it brightened again. I'll skip the explanation of how we finally unlocked its identity, but we now believe that Wormwood—that is, 2004 JU04— is a planet killer. An asteroid the size of Africa that is accompanied by six companions. These orbit it like miniature moons, and this small system bumps into other astronomical bodies along the way, which causes shifts in the overall trajectory. As Ben Brandeis will verify, the speed at which this system is traveling currently puts it on a close flyby with Earth in less than three years. As of this morning the probability that the Wormwood system will hit us has risen to as high as 69 percent, depending on which scientist you ask, but the bottom line is this: we have to prepare for it because there is no way we can move this planet to get out of its way."

Back in Marionville, Amelia McCutcheon stood in their church's food pantry, taking inventory of the food and water the committee had collected during a recent community drive. Her best friend and neighbor, Katy Kuppler, stood on an aluminum ladder, counting boxes of cherry fruit drink. "That's six dozen in all," she said as she descended the metal steps. The forty-nine-year-old homemaker brushed dust from her hands. "We'll need to give it all a good cleaning before we open to the public next week."

Amelia entered the tally of drink cases and closed the thick ledger. "I'll enter this into my computer program later, but it looks like the drive was a big success. You know, Mac's never been one to emphasize becoming one of those end-time preppers, but he called me last Friday and suggested we start storing up nonperishables in our basement. He thought our idea for a food pantry was a really good one. Is Lisa stopping by tomorrow to help?"

"I doubt it," Kuppler replied. "You know how she is, Amelia. She thinks folks need to buck up on their own. Lisa's generally good-hearted, but she's grown sort of odd about anyone who's down on their luck lately. I can't think what's gotten into her."

"'The love of many will wax cold,'" Amelia quoted.

"Huh?" her friend asked as she searched through a canvas handbag for a hand sanitizer.

"It's a verse in Matthew 24," Amelia answered. "Mac quotes it all the time."

Her friend laughed and snapped the leather closure on the bag. "Your husband quotes a lot of things, but I can't make heads or tails of most of it. Way over my head. Pastor Gibbons said most of Pete's books read like a doctoral thesis."

Amelia sighed. She'd heard this from so many of her friends, but Peter McCutcheon's warnings couldn't be clearer—or so she thought. Amelia had only a high school education, but even she had no difficulty comprehending Peter's works. He did use specific, often scientific phrases in his writing, but he always made them clear in the summaries of each chapter. Bullet point by bullet point. He'd even written teachers' guides for use in churches, but still most pastors ignored his warnings that the seals of Revelation were opening and that something awful was about to happen to the planet.

She thought of their most recent phone conversation. Since being reactivated with the air force, Mac had shuttled back and forth between Colorado and Washington, stopping in Indiana once to pack, but they talked every night by phone or video chat. As with his former job with the Force, Peter rarely spoke of his mission, but he sometimes offered hints, and these seemed tinged with great concern. Amelia had an intuition that her husband was trying to tell her something important, and she'd asked God to help her with discernment, but she'd come up empty.

Until last night.

Pete called at his usual time, quarter past ten, and they shared tales of the day's activities. Amelia mentioned news reports about a space probe and rumors about government crackdowns in China and South Africa. As always, Pete put her mind at ease by assuring her that God was still in control. Then he said something that struck Amelia as strange. He asked Amelia if she'd consider moving to Colorado Springs. Even when he was stationed there, Pete never once suggested his wife leave her childhood home or her parents. Their children had grown up in Marionville, and now their grandchildren would do the same. Amelia's parents were aging, and her mother's recent diagnosis of Alzheimer's made it difficult to think about relocating. What was Mac trying to tell her?

"You're certainly far away," Katy Kuppler noted, the comment jerking Amelia's thoughts back to the present. "I hope I didn't upset you by what I said about Pete's books. You know me, always opening my mouth and chewing on my size nine foot!"

"No, that's not it," Amelia answered. "I guess I'm tired. I didn't sleep all that well last night. I might stop by the nursing home on my way to the market and take a look at the Memory Unit. I've volunteered there often enough, but I want to find out the costs, you know? For Mom."

"Yeah," Kuppler said with a sigh. "Look, Amelia, I'm really sorry. I should just keep my opinions to myself. I wish my brain could process the stuff Mac writes, but it's a wasteland up here," she added, pointing to her temple. "Ask Ed! He's been telling me since the day we got married that I couldn't find my way out of a parking lot without his special brand of HPS."

"HPS? Don't you mean GPS?"

Kuppler laughed as they walked to the door. "Husband positioning system," she explained. "Come on. I'll go to the nursing home with you. And then we'll stop by the ice cream store. My treat."

Several long months passed, and the Wonderland Group settled into a routine of meetings punctuated by helicopter rides to a variety of locations—most underground. The core members kept in contact with one another through a secure communications line, but even then their conversations seldom used anything other than general phrases. Code words supplied to all the members allowed them to convey quick details without compromising their secret mission. Ben Brandeis had taught them his own ciphers, but only Mac and Dr. Stone used them. He and McCutcheon had grown quite close in that six-month period, and the older man looked upon the British genius as an adopted son.

It was in late November when a newcomer joined their conclave. A jocular fellow named Steve Campion, head of a Silicon Valley conglomerate, seeded with In-Q-Tel funds. Campion's wildly popular social media sites had benefited from a rogue computer virus that suddenly crashed all the servers of its biggest competitor a few weeks before the 2020 elections. Now, as the favorite choice for social media interactions, Campion's ChampChat served up news, games, and streamed content to half the world's mobile users. As a member of Wonderland, the thirty-three-year-old entrepreneur had been tasked by his government investors to use the social media company to shape public opinion and control any Wormwood leaks.

Something about the billionaire irritated McCutcheon, but he wanted to believe the best of people. Dr. Stone commenced their full meeting with her usual dry delivery.

"I need not repeat my admonition that our agenda and conversations remain top secret," the presidential adviser began. "However, there is one recent news article that requires our immediate attention. I refer, of course, to the malicious article in an uncooperative British tabloid about Wormwood. Now, I am perfectly aware that this name is biblical; therefore, we cannot conclude that there is a definitive leak from our inner circle, but I'd be remiss if I did not ask the question. I say this only because either the writer guessed that our governments are planning for citywide evacuations, or else someone informed him. It need not be said that if I ever learn it is the latter, the person responsible for the leak will lose his or her place in the refuge city."

A hand went up.

"Yes, Mr. Campion?"

"Refuge city? It's the first time I've heard that phrase in an official context, but I've noticed it trending on my sites."

"Really? Since when?" she asked him.

"Beginning a few days ago. I have an AI algorithm that collates all the pertinent trends and hashtags into discreet files, and these are delivered to me each morning with my cinnamon latte. Monday I found a file marked with the header #CitiesofRefuge. Do tell me that your people aren't the ones posting about this!"

Ben Brandeis raised a hand. "May I, ma'am?"

"Yes, of course, Ben."

He stood. During the previous six months, the young man had earned the respect of every member of the group, even the crotchety Secretary Gilmore, and everyone paused to listen.

"I used to spend a lot of time on social media, but lately I just haven't had enough hours in the day to hang out, you know? But I do try to stay on top of what's happening on the sites popular with my age group. No offense, Mr. Campion, but your site's already losing a lot of my friends because of all the political ads."

"Son, we don't run ads," Campion said proudly.

"Not ones that you label as ads, no," the teen answered without hesitation, "but the result is the same. I know more about coding than you've forgotten, sir, and I can recognize algorithms and bots in seconds. I noticed that *refuge* hashtag last week on one of the gaming sites, and it's already gaining steam. I know you've all tried really hard to keep a lid on what we're facing, but shouldn't people be given a choice? Have a chance to stock their pantries or maybe do what Major McCutcheon calls getting 'right with God'? Who knows? If the rebels in the Middle East and Africa knew about an incoming planet killer, they might actually stop fighting and try to find a way to help each other! Do we have to keep doing what we've always done out of habit?"

Gilmore cleared his throat and chuckled pridefully. "Typical for your generation, young man. You think peace is just around the corner, if we old fogies would just get out of the way, eh? Son, it doesn't work like that. We control the story, we control the access, and we can control the result. These refuge cities only hold a million people each, and there are only ten of them. Ten million human beings—period. That is all we can save. Better the fools in the Middle East keep killing each other off and reduce the tally of those that'll face this Wormwood killer. Fate is fate; that's a fact."

Mac held up his hand, angry at the insensitive politician's callous attitude. "Is it, sir? I mean, is this why we've been working overtime? To rescue ten million people? Look, the Bible says—"

"Do not start quoting the Bible at me again, Major!" shouted Gilmore. "I'm a Christian as well, but I fail to see how your so-called eschatology can help us now. If you think that divine intervention is comin' out o' the heavens, then I suggest you remove your name from the refuge cities list and depend upon the rapture or some other miracle to save you! All those prophecies were fulfilled centuries ago. They have no place at a table of rational men!"

"If you're a Christian, sir, then why do you fear God's Word?" McCutcheon answered patiently. "Suppose the Lord chose to intervene and save a hundred million, wouldn't you want that?"

"Of course I would," the politician replied proudly.

"Two hundred million?" the major asked.

"Sure. Even three or four or five, but that is simply not going to happen. There are ten cities in ten countries. Ten million people. And we need to fill them with the best of the best. That's the only reason that you and Ben are on the list, if you must know. Of course we've added your immediate families out of a courtesy— in gratitude for your expertise. Unlike your absentee God, our governments have provided safe havens."

"*My* God?" Mac echoed in shock. "Why is He *my* God? You call yourself a Christian, yet you do not call Him *your* God too?"

The aging politician cleared his throat, his shifty eyes casting about the room. "O' course He's my God, certainly. I merely try to make a point about your radical brand of Christianity. You're an outlier, Major. An aberration to the mainstream. Look here, let's just table this discussion until later, shall we? We need to let Dr. Stone lead the meeting. This isn't church, you know. It's a serious meeting of sensible men and women."

Mac couldn't let it go, and he found himself losing his temper. "By that do you imply that a church body lacks sense? I am a passionate believer in Christ, and I consider myself a very rational, reasonable man. If you want to go toe to toe on philosophy or science, I'd be happy to debate you, Mr. Secretary!"

"Now, see here—"

"No, sir, you need to see! I seriously doubt that you've ever read any of my books, but have you actually read *the* Book, the Bible? If you had, then you'd know what I've been saying for years is true. The world as we know it is about to change. If Wormwood is on its way, as we suspect, then that means the entire Revelation scenario is beginning. Maybe it's already begun, but no matter which, Jesus Christ *is coming back*. And that means the man of sin is coming as well. In fact he may already be here!"

Dr. Stone rose, her face impassive as usual. "I commend your passion, Major McCutcheon, but we cannot settle this argument at this meeting. I suggest you and the good secretary make plans to exchange your opinions in private. For now let me assure you all that the world's governments have not been idle. I've brought with me this afternoon a provisional document that reveals a new technology that may actually save us all and negate the need for refuge cities. Tomorrow we shall discuss this device and how it might be implemented at length, but I shall explain it briefly for now. You will each receive a copy of this proposal for private reading."

"You say it's a new technology?" asked General Maximilian Jarvis, head of NATO. "Why haven't I heard of this?"

"No one outside of CERN has heard of it, General. Not until now.

As I said, tomorrow we shall discuss the precise details, but the device utilizes a newly discovered entanglement property to fire time packets into the path of Wormwood."

McCutcheon actually laughed. "Time packets? What science fiction novel did CERN's people take that one from? To my knowledge, such a thing isn't even theoretically possible, much less practical!"

Dr. Holmes raised a hand. "Mac, I appreciate your amusement, and I'll admit that I had the same reaction when Eric Lindor called me about it, but the science does have merit. Ask Ben if you don't believe me. The theory's based on his equations."

McCutcheon stared at the teenager. "Your equations? Ben, you've not said one word about any of this to me. Why?"

"It isn't because I'm trying to keep you in the dark, sir," he told his American mentor. "It's just that a lot of what I theorize is sort of 'out there,' you know? It always makes sense to me, but most of our team's mathematicians and quantum physicists told me that I can't use entanglement as a constant. But that's not true. You *can*. And if you do, this can actually cause time to reverse."

McCutcheon sighed. He'd seen many of the teen's complex and beautiful equations, and though he considered them brilliant, the aging scientist found them unsettling. If Ben was right, then the nexus of dark matter with quantum entanglement opened a portal to a reality without any foundation at all. A reality that denied the need of any Creator God.

Wonderland. That's where they'd all landed. Down a deep, dark rabbit hole. Through a looking glass made of quarks and bosons and infinite impossibilities. The single ray of hope in this very dark place was found between Genesis 1:1 and Revelation 22:21. God's Word.

Mac thanked God for putting him in this place at this time. Isaiah 42:23 entered his thoughts: "Who among you will give ear to this? Who will hearken and hear for the time to come?"

"Maybe this is why I'm here," he thought.

Chapter 2

EYES ON THE SKIES

AFTER READING CHAPTER 1's elaborate story, the reader may sigh a breath of relief, realizing there is no real cause for alarm, as the narrative is purely fictional.

Or is it?

I intend to share significant reasons why I have come to believe undisclosed facts may be stranger—and scarier—than my opening fictional narrative. This should chill you to the bone.

And yes, this involves what I believe to be a cover-up of the highest order by national space agencies, including NASA.

Furthermore, I am not alone in my conspiratorial thinking.

In a recent peer-reviewed paper, "An Empirical Examination

of WISE/NEOWISE [Near Earth Object Wide-Field Infrared Survey Explorer] Asteroid Analysis and Results," physicist Nathan Myhrvold—former chief strategist and chief technology officer at Microsoft and a true polymath and working scientist who has published original research in paleobiology, climatology, and astronomy and who holds over 850 US patents issued to his corporation and its affiliates (*The Economist* once described Myhrvold as "Bill Gates's 'second brain'" and pointed out that most years he is on the "list of the world's 100 greatest thinkers"[1])—refutes asteroid data from NASA as suffering "from systematic errors and inconsistencies" regarding potentially deadly NEOs.[2]

Just short of whistle-blowing, Myhrvold goes on to charge NASA with deliberately misreporting the number of threats by near-Earth objects being tracked by their NEOWISE project, accusing scientists at the space agency of behaving "extremely deceptively" with deliberate "scientific misconduct" in a cover-up of very real and potentially imminent space threats.[3]

"The very NASA managers who should have been supervising the project were more interested in protecting it from scrutiny," he writes before adding:

> The issues I am calling misconduct in the NEOWISE papers were not inadvertent. They appear to have been deliberate choices made repeatedly by the NEOWISE team over a long period of time. These actions have caused the astronomical community to work under the false belief that the NEOWISE results are more accurate (have smaller errors) than the evidence warrants.[4]

Myhrvold pointed out:

> NEOWISE results were obtained by the application of 10 different modeling methods, many of which are not adequately explained or even defined, to 12 different combinations of

WISE band data. More than half of NEOWISE results are based on a single band of data. The majority of curve fits to the data in the NEOWISE results are of poor quality, frequently missing most or all of the data points on which they are based. Complete misses occur for about 30% of single-band results, and among the results derived from the most common multiple-band combinations, about 43% miss all data points in at least one band. The NEOWISE data analysis relies on assumptions that are in many cases inconsistent with each other.... After removing the exact matches and adding additional ROS [radar, occultation, or spacecraft] results, I find that the accuracy of diameter estimates for NEOWISE results depends strongly on the choice of data bands and on which of the 10 models was used. I show that systematic errors in the diameter estimates are much larger than previously described and range from −5% to +23%. In addition, random errors range from − 15% to + 19% when all four *WISE* bands were used, and from −39% to +57% in cases employing only the W2 band.[5]

In other words, Myhrvold's findings suggest the largest database in the world (with more data than all other sources combined) of information detailing diameter, albedo, and other properties of approximately 164,000 asteroids is suffering from intentionally manipulated information at worst and inadequate analysis at best, with the net result being that the public is being kept in the dark regarding... *what?*

As detailed later in this book, recent concerns voiced by other experts agree with mine and Myhrvold's and stem from variations in the formula for pi, which could mean a difference in the calculated trajectory of the soaring masses in space. If an asteroid or meteor in question is described with phrases such as "skim past" when its orbit near Earth is referenced, small errors in calculation could equal big consequences for our planet. For example, one mathematician, Harry Lear, recently asserted the claim that the formula used to calculate the trajectory of Apophis (a massive asteroid studied later in this work) by NASA scientists—pi = 3.141592654...—should have been 3.144605511..., with the golden ratio number calculated as 1.618033989.... According to Lear, this miscalculation leaves the space agency's numbers "off by 901,434 km too short for Earth and 831,517 km for Apophis,"[6] which Lear implies could send Apophis crashing into Earth in less than ten years from now, on April 13, 2029.[7] Lear has sent an open letter to President Trump and US government scientists, begging them to cross-check these calculations immediately, even though he ends his dispatch with an ominous admonition that we may already be out of time.

More recently NASA employee Robert Frost said the best thing governments could do if the mathematicians and scientists mentioned here are correct is tell the public to "hunker down,"[8] if indeed it turns out an asteroid capable of destroying life on Earth (such as the one that many scientists believe wiped out the dinosaurs) is disclosed by astronomers as heading our way. (Later I will go into why such an announcement has not or would not be made any sooner than necessary even if NASA privately confirmed the risk as real, which frankly it may have already done.)

In an *Express* newspaper interview with Frost, the facts behind his "hunker down" comments were explained thus:

> Despite talks of using nuclear weapons to obliterate asteroids or rockets to push them out of their path, NASA admits that if an asteroid got too close, it would be too late to save the planet. NASA said on its website: "An asteroid on a trajectory to impact Earth could not be shot down in the last few minutes or even hours before impact. No known weapon system could stop the mass because of the velocity at which it travels—an average of 12 miles per second."[9]

For the most part the public is unaware that 70 percent of such space threats (hundreds of thousands) are undetected and will remain so for the time being, partly due to NASA not approving the development of a full space-based telescope (such as one called NEOCam, which would vastly improve astronomers' opportunities to detect potentially dangerous NEOs). Even as I'm writing this chapter, the European Space Agency and European Southern Observatory confirmed that a twenty- to fifty-meter asteroid, 2006 QV89, which had a one-in-seven-thousand chance (an extraordinarily large risk assessment) of impacting Earth on September 9, 2019, "is not on a collision course this year."[10] Another "city-killing" asteroid NASA "didn't see coming" just zipped by Earth in July 2019. This asteroid—called 2019 OK—was traveling at 15 miles per second and came closer to Earth than our moon.[11] But asteroids don't all change in our favor. While I was writing this book, a press release from the University of Hawaii came across my desk, describing an incoming asteroid about four meters in diameter, which another article noted is "roughly the size of a small car,"[12] that astronomers only discovered as it entered our atmosphere June 22, 2019. And who can forget the almost sixty-five-foot-wide asteroid traveling at nearly 43,000 miles per hour that exploded in the atmosphere over Chelyabinsk, Russia, in 2013. Nobody saw that one coming either.

But the real apocalyptic risks are far more serious than fast-traveling, car-sized objects.

Imagine—as I describe more fully later in this book—walking out in the evening just a few years from now and noticing what looks like a horned, fiery serpent, thousands of feet wide, plunging through the heavens toward Earth at an incomprehensible speed. This terrifying monster seems to swim across the sky, past the stars, descending closer to Earth until, making contact, it plunges into the ocean, its massive form sending a sequence of tsunamis measuring six hundred feet in height slamming into the coastal terrain of regions across nearly half the world, infusing the atmosphere with scorched particles of aerosol and vapor. The resulting blistering culmination of moisture and extreme heat in Earth's atmosphere subsequently combusts into a series of high-velocity hurricanes, which turn their deadly gaze upon the yet unaffected hemisphere of the world. So much debris is released as a by-product of the initial impact and consequent devastation that for about a week darkness covers the sky worldwide as the entire landscape is pounded by hurricanes and similar atmospheric annihilation. By the time the waters finally settle, the storms subside, and the sky grows clear, most of the life on Earth is dead. According to Los Alamos National Laboratory environmental archaeologist Bruce Masse, this is precisely what those alive on Earth in the time of Noah saw, and *this* is how the deluge of that time took place.[13]

Need I remind you what Jesus told His disciples: "But as the days of Noah were, so shall also the coming of the Son of man be" (Matt. 24:37).

IS APOPHIS BIBLICAL WORMWOOD?

With that in mind, consider how in June 2004 astronomers at the Kitt Peak National Observatory detected a sizable asteroid heading in the direction of Earth. Subsequent efforts made later in 2004 by a team at the Siding Spring Survey in Australia identified the asteroid again. The next year, the team that discovered the asteroid named it Apophis (after the ancient Egyptian spirit of evil, darkness, and destruction, a malevolent force that cannot be stopped, according to legend). Immediately the possibility of a collision of the celestial body with our planet became a focus of calculation and preparation on behalf of preventive efforts by experts across the world.

The enormous asteroid Apophis will reportedly pass disturbingly close to Earth on April 13, 2029 (approximately nine years from the publishing of this book), according to NASA's website.[14] In fact

NASA admits Apophis in 2029 will be so close to Earth that it will "put some of our orbiting satellites in peril" and even be visible in the daytime sky.[15]

Assuming for the moment that Apophis is biblical Wormwood (I'm not certifying that or setting dates here) and that 2029 would thus represent a period sometime around the middle of the great tribulation period when the trumpet judgments begin, Monday, October 13, 2025 (April 13, 2029, minus three and a half years) would be the approximate start date of the dreaded seven years of tribulation foreseen in Scripture (see Matthew 24:21; Revelation 7:14; and Daniel 12:1).

For evangelical dispensationalists (and some Catholic prophecy believers) this timing may seem an ominous sign that a rapture of the church is soon to occur (the eschatological event when all true Christians who are alive will be transformed into glorious bodies in an instant and joined by the resurrection of dead believers, who ascend with them into heaven). Depending on one's particular position, this would place the last possible date for a pre-tribulation rapture happening sometime around October 13, 2025 (approximately six years from the publishing of this book), while for pre-wrath rapture believers this equals a departure time from Earth of just before 2029, and for a post-tribulation rapture, a taking-up/catching-away near Wednesday, October 13, 2032.

Bible prophecy notwithstanding, hopefully Apophis *will* pass just shy of Earth in those few years from now,[16] because the alternative truly represents a biblical-proportion risk to life on Earth. Measuring 370 meters across[17] and weighing an estimated twenty million metric tons,[18] Apophis is traveling at 28,000 miles an hour,[19] a mind-bending mass and potential inertia velocity encounter most people cannot begin to fathom. And while some experts have publicly stated that they do not, as of this time, expect the celestial mountain to make direct impact with our planet, Apophis' current categorization as an NEO, or near-Earth object,[20] has caused the same experts to remain watchful of the body's orbit, as that could change. Indeed, behind-the-scenes hypothetical plans to deflect the danger are under discussion because scientists see this upcoming "celestial event" as not only a rare opportunity for study and observation of a heavenly monster but also the possibility that they—as well as the rest of the world—may be spectators to a catastrophic collision with this planet in 2029. This is because along with the admitted closeness of Apophis' upcoming orbit comes the risk that an unforeseen event such as that portrayed in the opening narrative of this book could somehow, even spontaneously, modify the course of the massive, twenty-million-ton asteroid, devastatingly hurling it straight toward this world.[21] So the eyes of all humanity will most definitely be on this sign "in the stars" (Luke 21:25–26) very soon.

Science writer Greg Bear of CNN not only agrees but elaborates on the consequences of Apophis colliding with our planet in less than a decade from now, admitting, "If it hit Earth, the impact would unleash a blast the equivalent of over a billion tons of TNT [easily causing] billions of deaths and months, if not years, of climate disruption."[22] Considering the implications of an asteroid this size, anticipated to orbit so closely to our planet, and bearing in mind the potentially devastating fallout of collision, more experts have joined the efforts to evaluate what occurrence might cause the massive body to deviate from its course in the direction of Earth. They consider how Apophis still has nearly a decade to travel among other asteroids, planets, stars, and additional elements in outer space that could come into contact with it, potentially altering its path like a deflected billiard ball and interfering with the trajectory that experts now charter for the massive interstellar "game over" eight ball. Perhaps this is the cause of the extreme dual polarity we see between media and scientists, with news stories on one side, attempting to satiate public fear, and preparedness/intervention meetings quietly taking place among members of NASA, expert astronomers, scientists, and even the most influential global political leaders.

And then there are those theologians and prophecy students such as myself who see possibilities of another factor in all of this: the coming inescapable fulfillment of Bible prophecy. Note the particular descriptions of the first four trumpets of Revelation 8:6–12 and their aftermath. If you were to ask a scientist to explain what these verses seem to be depicting, he would tell you these details very much match the sequence of either a binary asteroid (two asteroids orbiting a common barycenter—the center of mass around which two or more bodies orbit) accompanied with smaller fragments, or the breakup of a larger asteroid into two main portions accompanied with tons of smaller debris as they enter Earth's atmosphere followed by impacts.

Note what these verses say in their exact original order:

> And the seven angels which had the seven trumpets prepared themselves to sound. The first angel sounded, and there followed *hail and fire* mingled with blood, and they were *cast upon the earth: and the third part of trees was burnt up, and all green grass was burnt up.*
>
> —REVELATION 8:6–7, EMPHASIS ADDED

Aside from the obvious correlation to the seventh plague of the exodus, "hail and fire" being cast upon Earth, scorching and setting ablaze fields and forests, easily matches the first, smaller portions of an incoming binary asteroid or debris detached from a larger space body upon impact with our atmosphere. Alternatively these minor meteor fragments could be debris caught in the gravitational pull of a single large asteroid that enters Earth's atmosphere ahead of the colossal rock.

Now, pay attention to what the very next two verses say:

> And the second angel sounded, and as it were *a great mountain burning with fire* was cast into the sea: and the third part of the sea became blood; and the third part of the creatures which were in the sea, and had life, died; and the third part of the ships were destroyed.
>
> —REVELATION 8:8–9, EMPHASIS ADDED

Immediately following the first trumpet with what I believe to be its smaller portions of red-hot fragmented asteroid materials, a picture emerges of a large burning mountain (exactly how I or ancients would convey seeing a sizable asteroid as it passes through the atmosphere toward this planet) impacting the sea, killing a third part of that ocean's life and wiping out ships with tidal activity. This appears to be the first of two larger parts of a massive disintegrating asteroid...or the first part of a binary space object.

Observe what happens directly following this, when the next trumpet sounds:

> And the third angel sounded, and there *fell a great star from heaven, burning as it were a lamp,* and it fell upon the third part of the rivers, and upon the fountains of waters; *and the name of the star is called Wormwood*: and the third part of the waters became wormwood; and many men died of the waters, because they were made bitter.
>
> —REVELATION 8:10–11, EMPHASIS ADDED

The third trumpet details what sounds like the second gigantic asteroid fragment "burning as it were a lamp" (again, exactly how John of Patmos would likely have explained a massive asteroid section blazing toward the world), which immediately impacts another part of Earth, causing tributaries to become polluted—including waters normally purified for drinking—and many people die as a result.

Finally, look at what this sequential description ends with:

> And the fourth angel sounded, *and the third part of the sun was smitten, and the third part of the moon, and the third part of the stars; so as the third part of them was darkened, and the day shone not for a third part of it, and the night likewise.*
> —REVELATION 8:12, EMPHASIS ADDED

Our hypothetical scientist, on reading these successive verses ending with verse 12, would explain how, as a result of the terrifying one-two punch of asteroid components plunging into the ocean and upon land, a sequence of tsunamis measuring hundreds of feet in the air would slam coastal terrains around the world, infusing the atmosphere with scorched aerosol particles, leading to extreme heat in the earth's atmosphere and a subsequent cascade of high-velocity hurricanes worldwide. Combined with voluminous debris hurled into the sky from the initial impact and volcanic eruption reactions, for about a week darkness would cover the sky, blacking out much of the light from heavenly bodies. In fact, as we detailed earlier, that is exactly what Los Alamos National Laboratory environmental archaeologist Bruce Masse says would unfold and in the same order as described in Revelation chapter 8.[23]

Of course this is only one scenario that makes sense to me and some scientists based on the descriptions in the Book of Revelation. Another possibility involves a second celestial body, officially called 2018 LF16, currently making its way through space "on a risk trajectory that might cause it to collide with Earth...[and] engaged in not one, but a staggering 62 different potential impact trajectories with our planet."[24]

The orbit of this asteroid has been monitored by NASA's Jet Propulsion Laboratory, which has calculated the mass' trajectory in an effort to mark its future course. Experts are saying this space rock, like Apophis, will safely pass Earth, but they admit that with so many potential collision points throughout the body's orbit, the likelihood of impact could increase greatly, should something unexpectedly alter its course.

If 2018 LF16 were to hit the earth, the damage stands to be only marginally less devastating than that of Apophis since the 2018 LF16 is slightly smaller. However, with such a higher numeric possibility of collision (the first of which is very soon—August 8, 2023), it remains an equally formidable threat despite its smaller mass. The asteroid is traveling through space at approximately 33,844 miles per hour and is estimated at about seven hundred feet across—"twice the height of the Statue of Liberty in New York and…four times as tall as Nelson's Column in Trafalgar Square,"[25] large enough to be considered a "country killer" were it to collide with Earth with a "cataclysmic collision [that would] generate the destructive power of a 57 megaton nuclear blast,"[26] creating a forceful explosion equal to that caused by the Tsar Bomb in 1961. (The Soviet RDS-202 hydrogen bomb, called the Tsar Bomb by Western nations, was the most powerful nuclear weapon ever detonated.) By comparison, this means that the devastation of the mass in question would be more "than 1,500 times that of the Hiroshima and Nagasaki bombs combined, and 10 times more powerful than all the munitions expended during World War II."[27]

Opposed to my preferred scenario earlier with the hypothetical scientist and evaluation of Revelation 8 trumpets 1–4 as a successive event, could 2018 LF16 alternatively be when the second angel sounds (possibly in the year 2023) and "as it were a great mountain burning with fire" is cast into the sea, followed six years later, in 2029, by the third trumpet and a second large asteroid called Wormwood? Will either of these near-date collisions thus provide fulfillment of the Book of Revelation's apocalyptic vision? A greater study of the devastation foretold in the Bible's final book, which we will analyze later, certainly outlines damage similar to that speculated by experts who are hard at work this very moment planning for just such a planetary crisis.

AND THEN THERE ARE THESE

While Apophis is, in my opinion, the most likely candidate to fulfill the Wormwood prophecy, NEOs skim past Earth all the time.

For example, NASA reported that on March 27, 2019, the NEO classified as asteroid 2019 EN was "taller than the Great Pyramid of Giza [and] shot past the Earth... [moving] at an incredible speed of more than 34,000mph."[28] This asteroid was only recently discovered but is estimated to have already had seven close approaches to Earth, with another pass projected to occur on June 6, 2035.[29]

Another asteroid, dubbed 2019 PP29, was recently discovered by Korean scientists utilizing the Korea Microlensing Telescope Network (KMTNet), which claims that the 525-foot body has a small percentage of risk of collision with our planet in either 2063 or 2069.[30]

Separately the second-highest placement of risk on the European Space Agency's "Risk List" for NEOs is the 1979 XB, a minor planet/asteroid that currently travels at over 90,000 kilometers per hour on a trajectory so unpredictable that its exact collision potential has yet to be precisely calculated but could occur anywhere between 2056 and 2113.[31] Experts advise that this mass is vulnerable to orbital variation, creating a higher risk than most NEOs, thus its placement as second on Europe's watch list.[32]

The 2010 RF12 is watched so closely that it currently presides over the ESA's lists. Moving through outer space at over 44,000 kilometers per hour, this smaller asteroid is sized at approximately seven meters across, and its potential impact brings with it ramifications slightly less threatening than those of the Chelyabinsk event of 2013.

Collision odds are predicted to be high at this time, with trajectories showing that this body will likely come forty times closer to our planet than our own moon travels. The asteroid is set to pass Earth in April of 2022, with calculated impact dates hovering near the end of this century.[33]

The 2000 SG344 is a unique asteroid in that it orbits the sun very closely to our planet, moving along its near-Earth orbit at over 40,000 kilometers per hour. As our planet orbits the sun in 365 days, this mass completes its cycle in 353 days, seemingly chasing our planet through space. This body is estimated at between thirty and forty meters across—close to twice the size of the meteor that caused the Chelyabinsk event—and could strike Earth within the next forty or so years.[34]

As awareness of these NEOs increases, the sheer volume at which they seem to hurtle past our world can be completely dumbfounding. On any particular day, multiple NEOs soar past Earth, some coming within close proximity and others clearing our planet at a comfortable distance. On April 6, 2019, alone, eight asteroids were reported to pass Earth, the smallest, asteroid 2019 GU1, being 30–66 feet in diameter but passing the closest to Earth of all eight passersby that day, clearing Earth at a distance of 391,007 miles away as it traveled. The largest of the visiting cosmic bodies on this particular day was the asteroid 2017 QP17, estimated at 1,098–2,455 feet in diameter, which fortunately granted our planet greater clearance as it orbited, passing at a distance of 36,818,044 miles from Earth.[35]

Another most ancient—and potentially volatile—threat that looms in Earth's future is the asteroid Bennu (named after an ancient Egyptian deity linked with the sun, creation, and rebirth, and formerly known as 1999 RQ36), which is thought to be one of the first elements to have formed in our solar system. The age-old mass, while stated to have low odds of impact, flies on a trajectory that holds seventy-eight prospective impact times with our planet on multiple dates between September 25, 2175, and the year 2199, the collision date with the highest potential being September 24, 2196.[36] While the official statement made by experts is that there is over a 99 percent chance that Bennu will pass Earth without incident,[37] rumors run rampant of a plan being devised by NASA to utilize asteroid mitigation measures upon the cosmic body to alleviate chances of collision.[38] Whether Bennu, estimated at 1,650 feet across (approximately one hundred feet wider than the height of the Empire State Building), is the actual object of mitigation efforts made by NASA remains to be officially disclosed, but the asteroid *is* currently the subject of the mission OSIRIS-REx (origins, spectral interpretation, resource identification, security, regolith explorer). Most of my readers are fully aware of Osiris as the Egyptian god of the dead and ruler of the underworld and his connection to the Great Seal of the United States and prophecies of his return in the last days connected to the spirit that will inhabit Antichrist. (One cannot help but wonder why international space agencies continue to name NEOs after ancient gods of destruction and what message they may thereby be extending.) NASA is obviously interested enough in this particular celestial member to chase a sample of its constitution. The OSIRIS-REx mission was launched on September 8, 2016, in pursuit of this aim and is set to return with its 2.1-ounce prize in 2023.[39] Calculated trajectories map Bennu's orbit as a safe pass-by of Earth rather than a collision, but NASA simultaneously states that the mass could also collide with Venus or could even potentially "burn up in the sun."[40] With such volatile would-be interferences looming in or near Bennu's orbit, the possibility of course alteration exists and could mean concern for Earth's safety. NASA cites many research-oriented causes for traveling 177,384,311 kilometers (for a total trip count of 2,585,737,262 kilometers as of July 1, 2019)[41] to *this specific asteroid* for information when many closer, more accessible options existed. However, NASA likewise acknowledges that Bennu's future proximity to Earth is "too-close-for-comfort,"[42] stating that "our descendants can use the data from OSIRIS-REx to determine how best to deflect any threatening asteroids that are found."[43]

DIDYMOS

In a recent announcement NASA indicated the SpaceX Falcon 9 rocket will soon carry what is known as the Double Asteroid Redirection Test (DART) planetary defense mission, which is set to launch from the Vandenberg Air Force Base in June of 2021 in a one-shot effort to interrupt the orbit of the 780-meter asteroid Didymos.[44] The asteroid, though not currently believed to be of threat to our planet, has been studied since 2015 and deemed by experts to be a good test object for a trial run of modern planetary defense and interceptive action toward the threat posed by cosmic bodies. The DART mission is charted to hit its target in September 2022, and experts are hopeful that this impact will serve to render valuable knowledge that should prove to be useful on future asteroids that show potential for collision with Earth.[45]

However, some individuals claim that such defense strategies could prove more difficult than initially assessed. Charles El Mir of Johns Hopkins University reminds experts that if nuclear bombs were sent into space to blast an asteroid apart, the gravitational force of larger space bodies may result in only temporary fragmentation, followed by pieces being drawn back together through the gravitational force of the asteroid, serving only to rearrange the shape of the debris surrounding a body without permanently redirecting or breaking up a large mass. He adds, "We used to believe that the larger the object, the more easily it would break, because bigger objects are more likely to have flaws. Our findings, however, show that asteroids are stronger than we used to think and require more energy to be completely shattered."[46]

36

An asteroid is defined by NASA as being a "rocky body orbiting the Sun."[47] As a result of having been forged beneath the heat of the sun during formation, asteroids contain different ratios of rock and metal, depending on their proximity to the sun during creation. As a result of these different balances of materials, asteroids are categorized as belonging to one of three compositions. The first is C-type asteroids, which are dark compositions of clay and silicate fragments. These are believed to be the oldest and most common variety. The second variety, S-type asteroids, consist of a blend of nickel-iron and silicate components. Thirdly, M-type asteroids consist of nickel-iron, being only metallic.[48]

While NASA has run many calculations on how the impact might change the trajectory of Didymos, it is NASA's stance that until the mission is carried out, it is not possible to calculate all variables that might affect the meteor's pathway. Elements such as "internal structure or composition," the asteroid's surface permeability, and other components could create deviations in results occurring in the actual event from those within simulations. Furthermore, the DART defense test run will carry a small device—the LICIACube—which will document the "impact and its aftermath…[acting as both] navigational system and camera to document the trip and mission," which was at one time projected to cost a whopping "$69 million altogether,"[49] although more recent estimates have returned at closer to $61 million.[50] Considering the fact that Didymos is not viewed as a threat to planet Earth, the price of such experimentation seems exorbitant, unless those conducting this test run see it as an investment toward protecting the planet from future, greater risk posed by similar celestial bodies.

Some add that the defense method posed by those who are coordinating efforts toward the DART-Didymos collision is to "'deliberately' crash a spacecraft moving 3.7 miles per second into it" in an effort to "strike [the] asteroid out of its orbit" in hopes of "demonstrating we can protect our planet from a future asteroid impact."[51] If this is true, it either lends a haphazard flavor to the endeavor or leaves one to wonder, Despite reports that currently moving NEOs are traveling at a safe trajectory for Earth, do experts know something that they are not telling the general public?

A PIECE OF THE WRONG PI?

While many experts, as noted earlier, maintain the optimistic attitude that all currently known NEOs will likely bypass Earth, it is obvious from the attention—and financial investment—directed toward planetary defense that the threat of possible collision is not taken lightly by those in the know. Additionally recent concerns voiced by some experts stem from variations in the formula for pi, as hinted at in this chapter, which could mean a difference in the calculated trajectory of certain asteroids. If the objects in question are projected to pass Earth with plenty of clearance, this may not seem like much of a problem, but when the objects in question are already being carefully observed and recognized by NASA as coming "close calls," small errors in calculation could equal big consequences for our planet. Mathematician Harry Lear's claims cited earlier are but one example.

And then there is *this* weirdness connected with the arrival of Apophis—the best candidate to fulfill the Wormwood prophecy, in my opinion—and pi. In 1981 a man named Billy Meier began to speak openly about a prediction he claimed was given to him by an "alien" he called Quetzal,[52] who allegedly told him of a "Red Meteor" that would collide with Earth in 2029, making impact "somewhere along the Tectonic Plate from the North Sea...to the Black Sea."[53] Although many of his claims (such as being given knowledge of the asteroid by an alien, or the existence of a high council of ETs in our galaxy[54]) have been considered fraudulent and "retrodictions" by some experts and members of the general population, the striking coincidence of Apophis' discovery, twenty-five-plus years after his prediction (if indeed evidence illustrates, as some claim, he foresaw 2029),[55] does raise one's curiosity. After all, the potential collision date along with likely location of impact (should this event occur) seem to hold almost uncanny similarity. Did an ancient entity indeed interact with Meier? We do know from Scripture that noncovenant peoples can sometimes nevertheless accurately predict the future. Here are just a couple of incredible examples of this from my 2017 book *Saboteurs*:

> Speaking of Nebuchadnezzar, his example also illustrates how in times past God sometimes used pagans to utter divine insights. An amazing case in point is when God chose to reveal a prophecy spanning from 605 BC through the Second Coming of Christ to the arrogant, narcissistic, idol-worshipping Nebuchadnezzar. Of course, it required God's holy servant, Daniel, to interpret the dream. Similarly, God used Balaam, a sorcerer hired by Balak, a Moabite king, who was exceedingly fearful of the encroaching multitude of Israelites. Accordingly, the king sent for Balaam, a darkened wizard who now lives in prophetic infamy (2 Peter 2:15; Jude 11; Revelation 2:14). Despite Balaam's incorrigible status, God used him to prophesy, "I shall see him, but not now: I shall behold him, but not nigh: there shall come a Star out of Jacob, and a Sceptre shall rise out of Israel" (Numbers 24:17). Ronald Allen, professor of Hebrew Scripture at Western Baptist Seminary, writes, "In agreement with many in the early church and in early Judaism, we believe this text speaks unmistakably of the coming of the Messiah. That this prophecy should come from one who was unworthy makes it all the more dramatic and startling." Thus, we see that God uses the most unlikely characters and situations to get His message across and work done. This Pethorian prophecy was well over one thousand

years before the birth of Christ and from a hostile source, yet it is probably what led the Magi to Bethlehem.[56]

Ironically, Meier not only predicted an Apophis arriving on the same date years before its discovery, similar to Lear's mathematical theory (mentioned earlier) he likewise holds that NASA's formula for pi needs to be recalculated. Both Meier and Lear also independently agree that should Apophis impact occur, disastrous aftermath would likely produce a crack in the Eurasian tectonic plate, potentially causing a split in the continent holding Europe and Asia and launching a series of volatile volcanic eruptions, which would subsequently unleash a massive quantity of poisonous sulfurous gases that would impact the waterways—as suggested in the Wormwood prophecy of Revelation chapter 8—in an uncontainable manner, creating a ripple effect throughout the entire world. As a by-product "hundreds of millions of people around the globe will die and our future crop seasons will be greatly impacted."[57]

When Apophis' orbit is recalculated using the alternative numbers (the alternate pi formula) suggested by Lear, Apophis' trajectory shows that it very well could crash into planet Earth on April 13, 2029, just as predicted by Meier in 1981.[58]

New York City Evacuated in Light of Impending Asteroid Collision (A Simulation!)

A simulation conducted by scientists and strategists from all over the globe took place in College Park, Maryland, in May of 2019.[59] The purpose of the gathering was to develop initial concepts around how such a global emergency might be handled by those in authority. At the five-day gathering the highly skilled partakers were given an asteroid scenario that hypothetically took place over a ten-year period, over which time these individuals had to attempt to avoid/respond to each emergency as it arose. The five-day simulation exercise included scenarios that fell into the following categories:

- "Day one—Discovery of large asteroid, will come close to Earth....

- Day two—Earth risk reaching unprecedented levels....

- Day three—It's heading to Denver. We need to stop it....

- Day four—Mission Successful—but now there's a new problem. Smaller asteroid is heading to the East Coast....

- Day five—'We did all we could, and it wasn't enough.'"[60]

Throughout this exercise the participants were faced with all foreseeable angles of the asteroid collision scenario. Beginning involvement highlighted the discovery and observation of an asteroid that was for that moment deemed to be of little or no risk to Earth. As collision possibility escalated, those involved were required to increase preventive measures in response while calculating possible damages and developing recovery strategies, should collision occur. As threat loomed at increasingly higher levels, intervention tactics rose as anticipation of devastation amplified. Deflection and nuclear disruption attempts were made, but these missions were only partially successful, leaving a major US city in the path of certain doom. As efforts were redirected from crisis aversion to damage control, new schemes emerged: those of evacuation and emergency preparation. At this point aftermath of the devastation was anticipated, and probable populace needs were estimated, along with potential by-products of such a disaster: from things such as meeting the tangible needs of the affected population to collective outflux, looting, rioting, and so on. At the end of the day, participants essentially determined there will be no stopping Wormwood.

Of course Apophis or any other currently observed NEO could indeed skim past Earth, leaving our planet unscathed. And to be sure, sensationalist predictions such as those made by Billy Meier can certainly serve to cast doubt on the viability of real, present danger. However, one has to believe that such experts as Paul Chodas of NASA's Center for Near-Earth Object Studies at Jet Propulsion Laboratory, Lindley Johnson of NASA's Planetary Defense Coordination Office, Leviticus Lewis of FEMA's Response Operations Division, and Rüdiger Jehn of the European Space Agency's Planetary Defence Office surely have better ways of spending their time than attending five-day hypothetical-scenario exercises...that is to say, unless the scenario is not considered by these highly qualified individuals to be so entirely hypothetical after all.

Furthermore, some experts, such as Iain McDonald of Cardiff University, explain that catastrophic events (such as asteroid collisions with Earth) occur periodically throughout history and cannot ever be completely ruled out as a future possibility. "There are always rocks flying through space. Inevitably one of these will hit us and it will have pretty dramatic effects."[61] McDonald was part of a crew that discovered an impact crater in Greenland beneath the ice in 2018—a find that is suggested to point to the culprit of the mysterious extinction of the ancient Clovis people.[62] At the International Academy of Astronautics Planetary Defense Conference earlier this year, NASA head James Bridenstine echoed: "We have to make sure that people understand that this is not about Hollywood, it's not about movies. This is about ultimately protecting the only planet we know right now to host life, and that is the planet Earth."[63] In fact Bridenstine explained that awareness regarding the possibility of asteroid collision with our planet has become a higher priority since the Chelyabinsk occurrence in 2013, an event equivalent to which, according to him, happens at least every sixty years.[64]

THE CHELYABINSK EVENT

The Chelyabinsk event took place when a twenty-meter (sixty-five-foot) meteor exploded in the atmosphere over a Russian city, causing a shock wave so profound that thousands of buildings were damaged and more than fifteen hundred people sought medical attention. "When it eventually exploded…it had…thirty times the energy of the atomic bomb at Hiroshima."[65] This was the largest recorded meteor collision in just slightly over a century, upstaged in recent years only by the devastating Tunguska event of 1908.

THE TUNGUSKA EVENT

Some currently assert that under worst-case-scenario circumstances a collision would create minimal damage. "As of 2014, the diameter of Apophis is estimated to be approximately 1,200 ft (370 meters). That's hardly a world-destroying size, but an asteroid of this size could obliterate a city."[66] However, statements such as these do not take into consideration a comparison made to similar events throughout history, which show evidence of much greater destruction than what would be categorized as minimal.

Take the Tunguska event of June 30, 1908, for example, when a blast theorized to have been caused by an asteroid estimated at approximately fifty to one hundred meters in diameter, much smaller than Apophis, collided with our planet, landing in Siberia. The impact was so dramatic that two thousand square kilometers of the Taiga forest was destroyed, "flattening about 80 million trees," and reports included accounts of the blast being felt by those in the closest city, over thirty-five miles away from the point of impact. In fact, the residents of the town, despite the distance from the impression, were able to feel the heat, were knocked off their feet, and even had windows bust as a result of the fallout. The blast was later said to have created "185 times more energy than the Hiroshima atomic bomb," with "seismic rubbles…observed as far away as the UK."[67]

Thankfully the location of the blast was a remote area, which is likely why, despite the fact that there were "hundreds of reindeer…reduced to charred carcasses," there were still no reports of human casualties as a result of the impact. An eyewitness to the event described it: "The sky was split in two, and high above the forest the whole northern part of the sky appeared covered with fire," and this sight was apparently followed by loud noises that sounded like guns firing. Furthermore, the nearby Lake Cheko is theorized to have been created as a by-product of the event. According to Luca Gasperini of the University of Bologna in Italy, "It [the lake] was formed after the impact, not from the main Tunguska body but of a fragment of the asteroid that was preserved by the explosion."[68] His theory maintains that the soft, swampy earth allowed a large meteoric fragment that originally landed on the surface of Earth to sink, creating a hole that morphed into the modern-day Cheko.

Notice the prophetic similarity to the eyewitness' account of this event. The individual described the sky as being "split in two,"[69] which seems eerily familiar to one who has studied prophetic Scripture:

> And the heaven departed as a scroll when it is rolled together; and every mountain and island were moved out of their places.
> —REVELATION 6:14

> And all the host of heaven shall be dissolved, and the heavens shall be rolled together as a scroll: and all their host shall fall

down, as the leaf falleth off from the vine, and as a falling fig
from the fig tree.

—ISAIAH 34:4

While the true source of this colossal event took years to confirm, speculation on the cause of the blast became rampant. Since
a subsequent search of the area did not render meteorite fragments,
the elusive source of the blast was the subject of much conjecture.
Those who supported the concept of meteor activity as a cause
claimed that the swampy surface of this area provided means
for the ensuing debris to sink. Others pointed to a comet. Since
comets are made of ice rather than rock, some individuals asserted
that the evidence would have melted into the earth's surface or
evaporated into the earth's atmosphere over the following weeks
or months. Further groups claimed that the blast was a result of
"matter and antimatter colliding," which could potentially result
in an explosion, while another crowd theorized that nuclear elements were at the core of the issue. Some even pointed to aliens as
the source.[70]

However, in 1958 researchers at the site discovered "tiny rem-
nants of silicate and magnetite in the soil," with high amounts
of nickel—all elements validating the case for a meteoric event.
Further analysis of the rocks found on this site, as recently as 2013,
confirmed meteoric origin.[71]

What would have happened if such a meteor had struck Earth
in a highly populated location? Or worse, if the hurling rock had
been much larger? When we consider the damage dealt during
the Tunguska event—forest destruction dealt over an area of two
thousand square kilometers (over twelve hundred square miles),
extinguishing eighty million trees and reducing "hundreds of reindeer...to charred carcasses" while possibly creating a hole in the
ground so large a new lake is formed, simultaneously breaking
windows and knocking people off their feet in a town thirty-five

miles away from the point of impact[72]—we are confronted with the catastrophe that Apophis would unleash with its approximately 370-meter body (potentially as large as seven times the size of that causing the Tunguska event of 1908), or the 2018 LF16 asteroid (estimated at over 210 meters,[73] or somewhere near four times the size of that causing the Tunguska event).

FULFILLMENT OF PROPHECY?

Could 2018 LF16, Apophis, or other objects currently classified as NEOs be harbingers of prophecies such as those in Joel 2:31–32 or Revelation? Could the asteroid Apophis or a similar celestial mass actually come close enough to do our planet harm, or worse, carry with it the judgment event that Scripture dubs "Wormwood"?

Many of us go about our lives making the assumption that highly skilled individuals are watching the skies on our behalf and creating defensive plans, while others believe the possibility of a devastating meteor striking within our lifetime is so low that we simply do not give it any worry. At the very least it is safe to say that for many learning about Apophis, 2018 LF16, Didymos, or even asteroids that have already collided with Earth, such as that of the Chelyabinsk event, this is likely the first real knowledge they have received on the subject.

And yet these enormous, forceful bodies are flying through space at speeds of up to 25 kilometers per second.[74] With so many thousands of asteroids, comets, and meteors zipping around in orbit, it may be overwhelming to even consider the implications. The enormous scope of this possible orbital conundrum, therefore, may be more relatable to a nonscientific reader in a different and more familiar context.

For example, consider how in rush-hour traffic in any large metropolitan area, thousands of folks are hurling down the interstate, driving 70 miles per hour in the same direction. Others, as they can, are entering and exiting the multilane expressway in this well-orchestrated symphony. All is well until the driver of an eighteen-wheeler attempts a lane change at an inopportune moment or a drunk driver enters the mix, driving erratically. Suddenly, unavoidable laws of physics come into play for everyone on the road.

Traffic goes sideways; brakes squeal. Every action does indeed have an equal and opposite reaction! Metal folds; airbags deploy as one vehicle grinds against another, mangling metal; and someone's car—or several people's vehicles—will never be the same again.

Magnified in a nonlinear progression, most readers can extrapolate the severity of a similar high-speed wreck taking shape well beyond our line of sight and far above us in space or skidding right into our hemisphere. Unlike with road traffic, an asteroid can do some serious damage just by coming too close to Earth's orbit. Those very real and problematic issues will be addressed in future chapters.

SUPERNATURALISM AND AGENTS OF END-TIME CHAOS

In addition to details about Apophis, its name draws some attention because the cosmic mass was named after the Egyptian god of chaos,[75] ironically also known as the Great Serpent, who, since he was also the enemy of the sun god Ra, existed in darkness. He is said by the *Ancient History Encyclopedia* to be "associated with earthquakes, thunder, darkness, storms, and death."[76] Considering the name from a point of irony, one could consider the reference to the Great Serpent who dwells in darkness as almost being a personified evil. Furthermore, if Apophis strikes the earth, chaos of the utmost kind would certainly ensue.

It could be that the implications stated in the previous paragraph contributed to the *very reason* that the name Apophis was chosen, and it might insinuate some at NASA suspect we are indeed on a trajectory with that chaos described in the Bible.

Another interesting element that raises suspicion on the part of some speculators stems from the fact that possible collision with Apophis would occur on Friday the 13th. As has been the case since the crucifixion of Christ, "the trepidation surrounding Friday the 13th is rooted in religious beliefs surrounding the thirteenth guest at the Last Supper—Judas, the apostle said to have betrayed Jesus—and the crucifixion of Jesus on a Friday, which was known as hangman's day and was already a source of anxiety."[77] Beyond the spiritual connotations of the events that launched the fear of Friday the 13th lies the general uneasiness people often have surrounding the number 13 itself, which is understandable, considering the fact that 13 has been noted throughout history for its association with dark forces.

In my previous work *Saboteurs*, I pointed out some of the nefarious connections made with the number 13, briefly revisited here:

A rudimentary search provides a lengthy list of superstitious connections to thirteen, many of which eventually trace back to the occult (if one is willing to dig that far back). Such a fear of this number exists that there is an actual, clinical phobia related to it. Triskaidekaphobia is derived from the Greek words *tris*, *kai*, *deka*, and *phobos*, translating literally "three and ten morbid fear," or "a morbid fear of three plus ten." The following are only a few interesting connections to thirteen:

- Most Wiccan covens are made up of thirteen members....

- On Friday the 13th, in October of the year 1307, King Philip IV of France had the Knights Templar arrested, most of whom were either tortured or executed immediately afterward....

- In Viking lore, the god Loki was the thirteenth in the order of the Norse pantheon. After his arranged murder of fellow god Balder, Loki was the thirteenth guest to appear at the funeral. (Some have said this is the origin of the superstition that if thirteen people gather together, one will perish in the coming year.)

- Apollo 13 was launched at 13:13:00 central standard time, and the oxygen tank exploded on April 13, 1970. [Note that April 13 is also the projected day of Apophis' next near-Earth orbit.]

- Early on in the development of our modern calendars, a year with thirteen full moons (which happens for approximately thirty-seven years out of a century) would throw off the balance of the scheduled church festivals, feasts, and events. As such, it was at the least considered an unlucky number and at times considered an ill omen.

- Many television shows and Hollywood films have plots surrounding the number thirteen in relation to the occult, paranormal activity, or supernatural phenomena (such as *13 Ghosts*, *The Thirteenth Floor*, *Friday the 13th*, and *Warehouse 13*, just to name a few).

- A surprising number of hotels and business centers have completely and superstitiously omitted the number thirteen from their flooring design, and as a result, the elevator buttons go straight from twelve to fourteen. (The same can be said for quite a few airlines as well; there is no "row 13.")...

- There are thirteen famines recorded in Scripture, which, as we all know, are times of terrible hunger and destitution for people of a country/territory (Genesis 12:10, Genesis 26:1, Genesis 41:54, Ruth 1:1, 2 Samuel 21:1, 1 Kings 18:1, 2 Kings 4:38, 2 Kings 7:4, 2 Kings 25:3, Nehemiah 5:3, Jeremiah 14:1, Luke 15:14, and Acts 11:28). In these events, the people were freed and fed only by a power higher than themselves, which was the Christian God. Consider, then, the fact

> that in the end times, the "savior" of such destitution will be the Antichrist of a World Order—before he ushers in the great destruction.

- Revelation chapter 13 is likely the most important and descriptive chapter of any book in the entire Bible concerning the Antichrist....

- In Gematria—the Assyro-Babylonian-Greek alpha-numeric code system frequently used in association to the Rosicrucians' beloved Kabbalah texts—the name "Satan" in both Hebrew and Greek are multiples of 13 (Hebrew 364, 13 x 28; Greek 2197, 13^3).[78]

Additionally Ethelbert W. Bullinger pointed out in his work *Number in Scripture* how the number 13 is permanently marked with the rebellious spirit of mankind:

> As to the significance of *thirteen*, all are aware that it has come down to us as a number of ill-omen. Many superstitions cluster around it, and various explanations are current concerning them.
>
> Unfortunately, those who go backwards to find a reason seldom go back far enough. The popular explanations do not, so far as we are aware, go further back than the Apostles. But we must go back to *the first occurrence* of the number thirteen in order to discover the key to its significance. It occurs first in Gen. xiv, 4, where we read "Twelve years they served Chedorlaomer, and the thirteenth year they REBELLED."
>
> Hence every occurrence of the number *thirteen*, and likewise of *every multiple* of it, stamps that with which it stands in connection with *rebellion, apostasy, defection, corruption, disintegration, revolution*, or some kindred idea.[79]

Considering this insight, it stands to reason that the number 13 could be a trigger element (a calendar date, perhaps?) for the judgment of God to be unleashed upon an insubordinate world. When the potential implications of Friday the 13th are coupled with Apophis' projected proximity to Earth and the possibly prophetic history of the asteroid—not to mention the potential destruction the mass stands to dish out should impact occur—the very mention of the asteroid named after the Egyptian god of chaos serves to incite misgiving on behalf of those who wait below to see what course it will follow as well as suspicion over whether world governments know—or suspect—more than they are admitting. And regardless of NASA's efforts to mitigate or obscure the situation, we know that someday an asteroid—or something *else*—of devastating implications possibly connected with the number 13 will collide with Earth, and when it comes, it will be carrying the curse of the judgment of the Lord Almighty along with it.

> The great day of the LORD is near, it is near, and hasteth greatly, even the voice of the day of the LORD: the mighty man shall cry there bitterly. That day is a day of wrath, a day of trouble and distress, a day of wasteness and desolation, a day of darkness and gloominess, a day of clouds and thick darkness, a day of the trumpet and alarm against the fenced cities, and against the high towers. And I will bring distress upon men, that they shall walk like blind men, because they have sinned against the LORD: and their blood shall be poured out as dust, and their flesh as the dung. Neither their silver nor their gold shall be able to deliver them in the day of the LORD's wrath; but the whole land shall be devoured by the fire of his jealousy: for he shall make even a speedy riddance of all them that dwell in the land.
>
> —ZEPHANIAH 1:14–18

Chapter 3

OUR DESTINY BEYOND EARTH

I F AN ASTEROID, comet, or other celestial threat were heading toward our planet, early detection and monitoring would be our first step toward defense. This stands to reason, as even early astronomer Galileo Galilei stated: "All truths are easy to understand once they are discovered; the point is to discover them."[1] Surely this man, of all people, would have understood how vast the mysteries of outer space can be, as one who designed and built such groundbreaking tools as telescopes and compasses, developed laws of physics that revolutionized the science of his time, made discoveries such as sunspots and some of Jupiter's moons, and generally overhauled the way mankind views the universe.[2]

Through the initiating of the US Space Force that will serve as part of the American military, one recurring theme has been that of early asteroid/comet detection. Surely any attempts to defend our planet against such flying objects will best be made when the earliest prevention planning and intervention can begin. But is that the only reason for a space force? Could the Book of Revelation's Wormwood be on the US government's mind? Never have more prophecy believers been part of an American president's inner circle than today. And I know for a fact—as described in my firsthand accounts in my work *Saboteurs*—they are "in his ear."

Certainly some of the dialogue relating to Trump's space force has been vague and for this reason has left room for much speculation as it pertains to the motivation and functions behind the development of such a military branch. When attempting to obtain an understanding of what has quickly been dubbed Trump's space force, it would seem that the best place to start is with the man himself.

PRESIDENT TRUMP'S SPACE FORCE

The idea of a space force isn't necessarily new, although President Trump has brought the concept to the forefront of the minds of the general public. To be sure, recent increased concerns regarding NEOs and potentially hazardous objects (PHOs) have certainly fed the interest, but the idea of maneuvering tactically from an orbital position goes back as far as the Cold War, when friction between the United States and the Soviet Union revealed to each side that having a foot in space may give it an advantage over its opponent. Later, in the early 1980s, Ronald Reagan initiated the Strategic Defense Initiative (dubbed Star Wars by criticizers), which would have been set into place for defensive purposes against the Soviet Union. Again in 2001 the issue of space defense was raised, by former Secretary of Defense Donald Rumsfeld, who concluded that the United States needed to increase the security surrounding its satellites, which by then had become essential to our daily lives and national security. In 2017 Rep. Mike Rogers again raised the question of the need for space defense.[3]

The air force has released a statement calling for an estimated thirteen billion dollars to be spent over the next five years to set up the US Space Force and Space Command, "a unified combatant command that will oversee space operations."[4] In addition to growing concern of NEO impact, authorities face an added burden that prompts the interest in increased security in space: the need for functionality of satellites. Modern innovations have increased the conveniences we enjoy in our daily lives and tightened national security but have placed great dependency on this equipment and its continued safekeeping. Up to now the air force has carried the burden of monitoring and defending American satellites in space,[5] but President Trump, like the individuals previously mentioned who raised the issue of national defense from a position in space, sees the need to respond to growing military demand in these areas by creating a space force and designating this endeavor as its own sixth branch of the military to serve in a separate capacity from the army, air force, marines, navy, and coast guard.[6]

The force's jurisdiction in space would involve addressing concerns of NEO or PHO jeopardy but would also operate more practical and full-time functions such as defending satellites utilizing systems that could "degrade, deny, disrupt, destroy, and manipulate adversary capabilities," should the need arise. Likewise some of the responsibility currently being carried by the air force—such as satellite positioning/tracking, space debris monitoring, and proprietary communication regarding prevention of satellite impact— would subsequently fall under the jurisdiction of this new military force.[7]

Furthermore, the United States would likely exercise its right under the Outer Space Treaty to place weapons in space, so long as they were not nuclear or weapons of mass destruction. The nature of such weapons could only be for defensive purposes, but as General John W. "Jay" Raymond of the Air Force Space Command stated, space is now being viewed as "a war-fighting domain and we need to treat it as such."[8]

While issues such as asteroid deflection are continually increasing in awareness among worldwide authorities, there are still some who would state that the risk is oversensationalized or that if a *true* risk were present, we would likely be unable to intervene on our own behalf. However, the introduction of satellite risk into the conversation always brings a tangible, *real* cause for alarm, which often rallies naysayers to confess that even aside from asteroid or comet danger, space defense is indeed becoming a timely concern. In consideration of such emergent issues, some authorities, even those in Congress, "have called for weapons that could destroy ballistic missiles from space."[9]

However, there has been some opposition to the creation of this division of the military—some to the general idea of a space force at all, and some to the concept of creating a sixth division of the armed forces. Many claim that while the air force has been shouldering these responsibilities, removing them from that department could result in bureaucratic and operational hassle for existing branches of the military that up until now have been successfully monitoring these procedures. For example, current personnel serving in space functions could be removed from the division in which they currently serve and find themselves funneled into this new, separate division of the military. Additionally the cost of developing this new branch has never been concretely estimated, with projected budgets from varying sources ranging from less than one hundred million dollars to between five billion and thirteen billion dollars.[10]

After some opposition from the Pentagon and Congress to creating a completely separate division of the military, Trump has acquiesced by launching a developmental stage of the space force beneath the umbrella of the air force for the time being.[11] It would take an act of Congress (literally) to create a sixth division of special forces within the American military. The last time this occurred was in 1947, when, under the leadership of Harry Truman, the air force was officially established.[12] However, on February 19, 2019, Trump signed Space Policy Directive 4,[13] a directive that orders the Pentagon to draw up legislation that would initiate the space force as an official, separate military branch operating under the jurisdiction of the air force. The president remains insistent that "our destiny beyond the Earth is not only a matter of national identity but a matter of national security."[14] Furthermore, he emphasizes: "America must be fully equipped to defend our vital interests. Our adversaries are training forces and developing technology to undermine our security in space."[15]

Nearly forty retired authorities in the fields of national military and intelligence have been vocal in their support for the establishment of a space force. Even as I write this, Congress deliberates on whether to acknowledge this force as a separate branch of the US armed forces.[16]

The strange X-37B

The placement of the space force within the jurisdiction of the air force is not surprising, since the mysterious X-37B space plane has been hosted from the Air Force Space Command, which has headquarters in Colorado but has launch locations in Florida and California as well.[17] The space drone, built by Boeing, is officially referred to as the Orbital Test Vehicle, or OTV. This vehicle was launched from the Kennedy Space Center in Florida for its fifth two-year orbit in September of 2017, a mission that the public has received nearly no information on, and the air force designates this craft to be "an experimental test program to demonstrate technologies for a reliable, reusable, unmanned space test platform."[18] Furthermore, the vehicle seems to come and go without public rallying, noted by a "landing that was announced to the public only via the sonic boom it created over Florida" when it returned from its fourth mission.[19]

The only official information released pertaining to the fifth mission (which the drone is still currently out on) is in regards to the Advanced Structurally Embedded Thermal Spreader, or ASETS-II,[20] set in place by the air force's research laboratory, which is stated to be currently testing "long term performance of an oscillating heat pipe (OHP) on orbit…to measure the initial on-orbit thermal performance, to measure long duration thermal performance, and to assess any lifetime degradation."[21] This vague statement could possibly mean that man-made equipment is being tested for long-range durability and functionality under conditions present in space, but it leaves much room for speculation as to the real *mission* of the X-37B. The sheer secrecy surrounding the craft's purposes lends mystery and thereby raises more questions than such vague descriptions as the previously mentioned one could ever begin to answer—such as: What is the *real* reason that authorities wish to develop a space force, and why, when much of the technology has been in place for decades, are they *now* beginning to pursue this effort so vigorously? What's behind the sudden rush?

Preparing for space traffic?

In addition to plans for making a space force, Trump likewise has declared that his plans for going forward will be carried out with the same assertive, motivated nature that the man imposes toward anything he sets his mind to: "We'll be setting aggressive timelines, challenging old ways of doing business, and we will be expecting real results," to which, he added: "I am instructing my administration to embrace the budding commercial space industry."

The president also states that a priority in this endeavor will be to modernize what he calls "'out-of-date' space regulations." The president has signed a directive for federal departments, agencies, and American industrial representatives to collaborate in an effort toward creating a "state of the art framework for space traffic management."[22]

The question raised by this motion becomes, What type of traffic are we truly expecting? Granted, the number of satellites currently placed in space will likely increase over coming years, and those that orbit at this moment perform tasks that, if interrupted, could compromise national security for the country of ownership (more on this later). Thus, for satellite traffic, additional updating and creation of legislation stands to reason. But are there additional types of space traffic anticipated in upcoming years? Current conversations seem to affirm these possibilities, citing potential new categories of such activity—possibilities such as commerce (including, but not limited to, asteroid mining), asteroid deflection and exploration, increased satellite activity, militia-related traffic, and even future colonization of locations such as the moon or Mars. In keeping with the ever-expanding list of new possibilities comes a perpetually mounting stack of potential risks, complications, and even vulnerabilities where national and planetary defense is concerned. As a result, authorities see cause for anticipating/resolving potential crises in advance by creating legislation to keep peace.

THE OUTER SPACE TREATY

In January of 1967, after a decade of deliberation and negotiations regarding issues such as military disarmament and enforceable territory between Earth and space, the Treaty on Principles Governing the Activities of States in the Exploration and Use of Outer Space, Including the Moon and Other Celestial Bodies (more commonly known as the Outer Space Treaty), was signed in Washington, London, and Moscow.[23]

The arms agreement established two vital points of settlement between all parties involved—it is agreed that 1) nuclear weapons or "weapons of mass destruction" are not to be set into Earth's orbit, on any celestial body, including the moon, or placed anywhere else in outer space; and (2) the territory of outer space must only be used for peaceful purposes, banning all celestial bodies (including, but not limited to, the moon) from use for military purposes, weapon testing, or tactical, military-related maneuvering.[24] Furthermore, the treaty was designed to "prevent space land grabs, similar to the territorial claims that plagued the exploration of Antarctica in the first half of the 1900s"; thus it established that the exploration of space should be an open and free endeavor to all nations and that property therein could not be claimed as territory belonging to any one country or entity.[25] Beyond this the document establishes the harmony under which exploration and conquest should take place and gives endowment for all astronauts to be given provisions or assistance in the case of distress or emergency, regardless of national origin or political affiliation.[26] Considering that this document was initiated over fifty years ago, it stands to reason that there would be need to update the terminology, add new legislation, and elaborate on elements that may necessitate such action as a result of recent discovery, modern military and technological innovations, or any other prospective scenarios that come to light as potentially needing to be addressed.

TECHNICALITIES OF THE SPACE FORCE'S OPERATIONS

As a by-product of the secrecy surrounding the X-37B, many people speculate there to be a hidden agenda behind the motivation for such a military endeavor. And to be sure, there are likely many elements to the undertaking that the general public has not been (nor may ever be) made aware of. Another thing that lends to the suspicion of cover-up or conspiracy is the fact that as of this time, the technical operations of a space force are still not clearly outlined. It could be said that this is partly due to the infancy of the concept of a space force, except that since the idea was introduced as far back as the 1980s, one wonders if the responsibilities and missions of this force are more developed than we are made aware of. It would certainly be a fair statement to say that the increasing dependency on satellites, alongside the perpetuity of growing technological advancements, likewise adds to already developing operations as situations arise, which could explain some of the vagueness regarding what the force would do. However, at this time there are some technical functions being outlined to the general public, and they bear mentioning here if one is to understand the pertinence of such a division of our armed forces. And yet many acknowledge that this particular angle of the discussion of the space force is "one of the foggier parts of the proposal... [and that] it's not completely clear what... enlistees would be doing."[27]

It has been stated already that a large portion of the space force's daily operations will be defending and monitoring our satellites. It is likely that systems will be set into place that will defend on two levels: 1) circumvent attempts to jam their signals, and 2) monitor and protect the satellites themselves from possible collisions with space debris, other satellites, or attacks such as missiles.[28] Essentially the official statement is that recruits will monitor space traffic. But if this were the entirety of the case, why design an entire additional branch of the US armed forces around this responsibility?

Efforts to map the entire (solar system) sky were made by a satellite system called Wide-field Infrared Survey Explorer (WISE), launched on the morning of December 14, 2009, from Vandenberg Air Force Base in California. This unmanned satellite orbited throughout space, using infrared light to collect pictures of all angles of the sky. These pictures are a vast resource of knowledge for those currently preparing for increased space traffic, providing images of galaxies, stars, and asteroids, and even such anomalies as black holes. The WISE satellite was semi-retired in 2011 when its voyage ended, but in 2013 it was sent out on a new mission: locating NEOs.[29] Thus the satellite was given the appropriate new name NEOWISE and was relaunched "to assist NASA's efforts to identify and characterize the population of near-Earth objects." Images collected and archived by this craft exceed ten million, and information gleaned as a result of this endeavor has assisted in the discovery of "158,000 asteroids, at thermal infrared wavelengths, including ~700 NEOs… [and] 34,000 new asteroids, 135 of which are NEOs." Beyond this the NEOWISE has discovered 21 new comets, while assisting in the tracking of more than 150 comets.[30] In addition to detection and tracking of various NEO bodies, the NEOWISE helps NASA configure the makeup of these bodies and calculate their size and trajectory.

Congress has requested that NASA locate and categorize at least 90 percent of NEOs over 140 meters in size by the year 2020 (more on this later). An 8.4-meter survey telescope called the Large Synoptic Survey Telescope (LSST) has been designed for just such a purpose, featuring a mirror the size of "a singles tennis court."[31] This invaluable tool will be used to record images of the sky by taking enough pictures to map the "entire southern hemisphere of sky every 3 nights for over ten years"[32] and is projected to pinpoint 75 percent of the NEOs that are 140 meters across or larger.[33] It utilizes a 3.2-gigapixel camera—the world's largest built for this kind of imagery, rendering images that offer coverage larger than "40 times the size of the full moon. LSST will provide a thousandfold increase in capability over current facilities."[34] The LSST is projected to capture "37 billion stars and galaxies," providing "10 million alerts, 1000 pairs of exposures, [and] 15 Terabytes of data" every single day.[35] As recently as June of 2019, this telescope was still being integrated with software and an auxiliary telescope (known as AuxTel) that will assist the telescope's information by measuring "atmospheric transmission during LSST operations."[36]

In order to increase from 75 percent identification and categorization of NEOs over 140 meters and reach the Congress-mandated 90 percent, the infrared space telescope called NEOCam has been designed, and it currently awaits approval for use by NASA after the design underwent fourteen years of fine-tuning. Because the infrared sensor collects thermal data to more accurately read the size of NEOs for better risk assessment, this will be more effective and empowering than a regular space telescope, which relies on the application of light and reflection to estimate size. It is anticipated that readings taken by NEOCam will provide reliable information and thus boost our defenses and ability to prepare. The equipment is tentatively set to launch in 2024, provided that approval and funding can be met by that time. Hopes are that the IMAP (Interstellar Mapping and Acceleration Probe) space probe will carry the NEOCam into orbit as soon as possible due to the cutting-edge upgrade this fixture offers those attempting to identify and categorize NEOs.[37]

The Double Asteroid Redirection Test (DART) mission was mentioned in the previous chapter, but it bears revisiting in greater detail, since at this point it is the best example of exactly how NASA and other experts around the world are proposing that we Earthers defend ourselves against potential collision. As mentioned already, the DART mission targets the asteroid named Didymos, which was given its name because it is the Greek word for *twin*, and this mass is actually made up of two smaller bodies, not just one large piece.[38]

The main bulk, known as Didymos A, is estimated at approximately 780 meters,[39] while the smaller portion, referred to as Didymos B (also dubbed the "Didymoon"[40]), is calculated to be about 160 meters. It is this smaller mass that NASA intends to impact with the DART mission. While experts have not confirmed the composition of the smaller portion as of yet,[41] they anticipate the impact's result to yield much knowledge about asteroid structure for future deflection efforts.

Recall that the Didymos asteroid is not currently considered a threat to Earth but makes a good example of the type of cosmic body that may someday need to be deflected, and so this trial run of NASA's plan and equipment has been scheduled for the unsuspecting orbital form. The objective of this endeavor is to impact Didymos with a projectile called a "kinetic impactor"[42] in an effort to alter the asteroid's course.

The logic for choosing a binary (made up of two main pieces) asteroid for such an endeavor is the fact that the Didymoon (Didymos B) orbits Didymos A, offering NASA the opportunity to target the smaller fragment and observe its response to the kinetic impactor, while it is believed that the orbit of the larger segment will be unaffected, causing the main mass of Didymos to continue in its current direction around the sun.[43] Thus the complete experiment should be observable, for all practical purposes, while banking on the strength of Didymos A's orbital pull as a sort of safety blanket from knocking the body completely off course.

The details of the mission seem simple enough: once it has launched, the DART craft will navigate its way to Didymos and, using "an on-board autonomous targeting system to aim itself at Didymos B," will point itself at the Didymoon and ram it at nearly 4 miles per second. Scientists will be watching the entire event and will be especially noting any orbital changes *between* the two fragments of the cosmic mass, which will yield information regarding future asteroid mitigation attempts conducted through kinetic impact. The goal is not to blast the asteroid into a completely different direction—having potentially disastrous future orbital results—but to lightly prod the body's trajectory onto a slightly different course that will deviate further from the "danger zone" over time.[44] Andrew Rivkin of Johns Hopkins University's Applied Physics Laboratory (APL) stated on this matter: "With DART, we want to understand the nature of asteroids by seeing how a representative body reacts when impacted, with an eye toward applying that knowledge if we are faced with the need to deflect an incoming object."[45]

The equipment for this undertaking is being designed, built, and managed by the Johns Hopkins APL.[46] The contract for the mission's launch was awarded to SpaceX in April of 2019, and the mission is scheduled for takeoff on a Falcon 9 craft in June of 2021.[47]

Of interest in addition to the mission itself is the fact that experts across the world will be watching for indications of success or failure. The European Space Agency has gone so far as to design what is being called a "companion mission," a probe that will follow the DART spacecraft and observe and report on the entire event. As experts work together to solve the issue of planetary defense, many prominent figures make statements inviting the collaboration of qualified individuals from all over the globe, each noting that cooperative capability surpasses the confines of national lines of separation.[48]

DAMIEN IWG

In June of 2018 the Detecting and Mitigating the Impact of Earth-bound Near-Earth Objects (DAMIEN) [NASA once more playing off how NEOs could be connected to the arrival of the Antichrist? DAMIEN is the Antichrist and the son of the devil in the fictional movie series The Omen] Interagency Working Group (IWG), working under the jurisdiction of the Committee on Homeland and National Security, in conjunction with NASA, released a document titled the "National Near-Earth Object

Preparedness Strategy and Action Plan" in accordance with the National Science and Technology Council, by which the Executive Office of the President endorsed a preparedness and response plan for NEO-related catastrophes. This document highlighted five main objectives for addressing, preparing for, and responding to NEO-related emergencies:

1. Increase the ability to discover, identify, track, and categorize NEOs.

2. Advance accuracy for projecting and simulating NEO characterization and course for improved decision-making abilities.

3. Cultivate strategies and technologies for NEO interference, including course deflection and disruption of bodies that pose a risk to Earth.

4. Encourage international collaboration regarding preventive and responsive strategies in regards to NEO risk.

5. Develop and practice NEO emergency response procedures and further advance action protocols.[49]

Within this document it is stated that NASA is currently tracking and cataloging near-Earth bodies that it deems potential risks. NASA estimates over three hundred thousand bodies measuring at least 40 meters in diameter pose a possible risk of impact that would be hard to determine more than a few days in advance, and there are approximately twenty-five thousand larger than 140 meters, one-third of which have been cataloged and due to their size are easier to detect in advance.[50] Collision with one of these larger bodies, however, could have the capability of devastating entire continents, since the energy detonated upon impact would be more than the power of any nuclear device ever tested, comparing to over sixty megatons of TNT.[51] Furthermore, this document acknowledges that Congress has requested that by the year 2020 NASA detect at least 90 percent of NEOs measuring over approximately 140 meters.[52]

Trump's administration seeks to bring agencies together for the purposes of NEO detecting, tracking, and cataloging, and characterizing the impact threat of these bodies while strategizing on possibilities for deflection, disruption, and even catastrophe response and recovery. Strategies are built around the objectives of assessing threat, choosing an appropriate path of action based on risk, carrying out said course of action, and crisis recovery.[53] Timelines for strategy plans include the two-year, two-to-five-year, five-to-ten-year, and ten-plus-year marks.[54]

This thorough document calls for improvement to the current ability to detect and catalog NEO threat by stating that current resources are available to "any existing national capabilities"[55] that might further current technology for this purpose, including opportunities to invest in improved telescope programs, further developing tracking abilities, and gaining ground as it pertains to NEO type characterization. Beyond these items, increased response speed and simulation scenario training were stated as desired outcomes.[56] Procedures for NEO deflection or disruption, still in developmental stages, are outlined in the text, along with conceptual elements regarding improvement for these areas.[57] Beyond the efforts to detect and possibly even intervene regarding NEO risk, the statement released from the Executive Office likewise clarifies that an essential part of risk mitigation is emergency preparation on the ground. Generally speaking, this article, alongside the fact that the space force is being formed, serves as proof that our government recognizes the risk posed by NEOs and is aggressively pursuing the initiation and expansion of a complete military-backed protocol for detecting, assessing, preventing, and responding to the aftermath of space-related national or global emergencies.

ASTEROID MINING

At first glance one may jump to the conclusion that asteroid mining is presented as an answer to the obstacle of limited resources here on Earth. For example, metals in short supply on this planet are said to be available in surplus among the stars. It is currently stated that "several privately funded space companies are locked in a race to claim the trillions of pounds worth of precious metals thought to exist in asteroids."[58] However, the sheer financial numbers involved immediately cast doubt on the long-term profitability of mining space's resources, should the end goal be merely returning the goods to Earth.

Furthermore, the current value of these precious metals to supposedly be mined from asteroids (the market price set for them here upon Earth) is currently derived from their rarity. However, as we all know, supply and demand are what regulate the fair market price for any particular good or service. So while some say that profits can be gained from iron, platinum, or gold that can be mined from asteroids, it is a shortsighted statement at best, a cover-up at worst. Economic authorities may project future profits at elevated prices—prices that are currently believable based on the existing demand vs. rarity of said supplies. However, if and when new, unlimited sources of these materials are tapped into, they will likely become more plentiful, which always drives the price of goods down. As values descend, investors may scramble to recap monies spent on expensive technology developed or purchased to obtain these materials in the first place, resulting in overfarming or flooding a particular market with the items of subject. Thus if asteroid mining is limited to collecting materials and bringing them back to Earth, the best-case result of asteroid-mining efforts could be depleted earthly values for such supplies and an obsolete, exorbitant fleet of asteroid-mining vehicles.

However, asteroid mining is broadened to include water in space as a vast resource. According to Planetary Resources, water is "the key resource in space... [enabling] the sustained presence of humanity in space... necessary for life support and [it] can even be refined into rocket propellant so that humanity can further explore the Solar System." These individuals assert that high-efficiency rocket propellant can be produced from water in space, allowing for the creation of "a space highway with fuel depots located at various points." Furthermore, many of those who wish to mine precious metals from asteroids state up front that they are not looking to bring the materials back to earth but rather are looking for local sources for said materials in order to support construction in space.[59] Additionally asteroids worth the expense of mining must be chosen carefully. Criteria being discussed at this time include the estimated mining value being no less than one billion dollars and the cosmic body being at least one kilometer in diameter, having a minimum platinum content estimate of at least ten parts per million, and having a travel rate of 4.5 kilometers per second. One other small detail must be noted: because of current technological limitation the only asteroids that can be reached for mining purposes from the earth are our NEOs. That's right—the asteroids that we are planning to tamper with are those traveling closest to our planet! When asteroids near Earth are analyzed using the previously mentioned criteria, we are only left with (at this time) approximately ten potentials that fit this description and are close enough to mine, meaning one of two things: 1) resources in space are more limited than anticipated, risking outrageous investments that could possibly yield no return, or 2) there is a different, longer-term plan that is anticipated to take us farther than our local corner of the solar system in search of treasures. Early phases of such endeavors could begin to materialize as early as 2020, when the Asteroid Mining Corporation could launch its prospecting satellite, should all go according to schedule.[60]

Recall that in chapter 2 it was mentioned that one reason Apophis has not been completely ruled out as a threat (meaning that experts continue to monitor its orbit) is the fact that should something interrupt its orbit, it could be deviated *toward the earth*. Why is it, then, that many of these same experts are not more alarmed by the concept of meddling with other NEOs orbiting near the earth, when one considers the fact that, by their own admission, a small interference could place our entire planet at risk? In light of such potentially devastating risk factors and possible financial fallout, one must find the true motivations and end-game goals behind asteroid mining suspect.

How much would such an industry cost to launch? In 2012 the Asteroid Capture and Return mission, accompanied by CEO of Planetary Resources Chris Lewicki and former astronaut Tom Jones, discussed the possibility of mining asteroids near Earth. The intended process would be done by selecting a suitable asteroid to target, moving it into lunar orbit where it could be tapped, and setting into place the necessary provisions for extracting its resources. In total the plan was estimated at $2.6 billion. Potential obstacles (possibly causing additional expenses) include the inability to locate a suitable asteroid within range that is of movable size and proximity, actual propulsion (or other relocation method) of said body once it is located and chosen, and the method and technology of *actually tapping* the resource once the asteroid has been moved into lunar orbit for harvest. The program reported: "Retrieving an asteroid for human exploration and exploitation would provide a new rationale for global achievement and inspiration. For the first time humanity would begin modification of the heavens for its benefit." While some authorities claim that such profits would go for the betterment of all mankind, one must wonder if this could possibly be true when others, likewise involved in studies regarding asteroid mining, make such assertions as "The future of space resource exploitation lies in the private sector."[61] Unfortunately there will be no real, *accurate* way of estimating the costs of such endeavors beforehand. By doing this, we will be stepping into uncharted territory. And not only does such uncharted territory bring unknown and incalculable *financial* risk; it likewise brings unforeseeable *physical* risk. What if, in an attempt to relocate an asteroid to the lunar orbit, we lost control of it, hurling it toward the earth? Does this seem impossible? How could we possibly know for certain when we can't even adequately estimate the financial cost of such an endeavor?

The lowest estimated cost of the upcoming DART mission to Didymos stands at $61 million.[62] When one considers the implications of this mission, similarities can be seen only to a certain point between the goals of each undertaking. DART will navigate toward Didymos and create a detonation in an effort to redirect the asteroid's orbit. Mining will have many of the same initial tasks: it will navigate toward the body and then have the capability of freeing the metals or resources within the asteroid, likely by detonation as well. However, this is where their similarities will end. DART will have equipment that records the event and transmits the information back to Earth. An asteroid-mining mission will have to have the means of collecting the elements it has extracted, a place to return said materials to, and a means of navigating and transporting these substances to their next location. Surely the cost of the Didymos/DART mission will be a fraction of what it will cost to set up asteroid mining. In light of this is it really cost-effective, or is there some other motivation to this effort?

Furthermore, it has been said that space is similar to the Wild West at this moment,[63] being an unclaimed, as of yet unlegislated territory, ripe for the picking to those first arriving on the scene, and lacking governing bodies that would regulate those who might attempt to generate commerce there—if the future of space-generated profits truly does lie "in the private sector."[64] If this is the case, then why, when it would cost no less than one hundred million dollars[65] to create a space force for us to use to defend ourselves against NEO asteroids, would we turn around and allow private organizations and corporations to tamper with things that may interrupt asteroids' orbits or place our planetary security in jeopardy for private monetary gains?

And yet one wonders if that is precisely what will happen when news emerges such as the announcement in June of 2019 that the International Space Station (ISS) will open up to more private companies looking for commercial opportunities, possibly as early as 2020. A lack of usage restrictions to these astronauts was called "unprecedented" by The Verge, which explained that these privileges, in addition to facility use, included "filming commercials or movies against the backdrop of space…[and allowing] private companies to buy time and space on the ISS for producing, marketing, or testing their products." It would even appear that NASA has given permission for these agents to utilize NASA astronauts as advisers, so long as their likenesses do not appear in private advertising. The statement was made in conjunction with a reminder to readers of NASA's request that members of the private space industry pass along any ideas for "habitats and modules that can be attached to the space station semi-permanently." The downside to this newfound open-door hospitality is that the quantity of visitors allowed per year is limited at this time, and the fees involved for necessary facilities are, as of now, exorbitant. These services include such life-supporting amenities as toilet systems, air, food, medical provisions, power, and other features that we take for granted here on Earth but that can cost as much as $35,000 per day on the ISS, according to NASA's chief financial officer, Jeff Dewit.[66] In addition to other amenities being made available on the space station, overnight docking ports will be accessible for companies that wish to utilize such private facilities for their own crafts. It's like a giant RV park in the sky.

It would seem that the more we discuss the real motivation behind space endeavors, the more a duality of purposes seems to surface: on one hand, there are the matters of planetary defense and national security; on the other, there seems to be an agenda of private gain and even eventual stabilized economy in space. One points to necessity, while one gestures toward colonization. And yet one is forced to wonder if the two purposes can coexist. Certainly many at this time would attempt to argue that the answer to this question is yes, and yet when we look around at our own planet, many will agree that we have not been good stewards of the world that we have already been entrusted with. When considering the potentially devastating possible consequences of tampering with elements such as NEOs or other components outside our world, while pondering the mess we've already made of this one, the reader should be compelled to ask the question "*Should* mankind be dabbling with outer space at all?"

SPACE TERRITORY TO BE USED FOR INTERNATIONAL—AND SPIRITUAL—WARFARE?

Many well-established figureheads within military realms insist that national security rides on the United States' ability to create and maintain a prominent presence in space, while others assert that the establishment of such manifestation will advance the odds of warfare within the jurisdiction of outer space. Even the president himself stated his own position on this matter: "Space is a war-fighting domain, just like the land, air, and sea."[67] Furthermore,

Trump said, "It is not enough to merely have an American presence in space. We must have an American dominance in space."[68]

Additionally influential individuals within agencies such as
the air force and even Mike Pence have been known to make the
assertion that China and Russia are already utilizing space as
a "war-fighting domain,"[69] and that they feel positively about the
American military taking this step to level the playing field. It did
not go unnoticed by American officials when China, in 2007, shot
its own satellite down in a sort of practice run that the New York
Times called "flexing muscle" and arms control experts called "a
troubling development that could foreshadow an antisatellite arms
race."[70] In fact China's actions caused many to claim that this
motion was a violation of the Outer Space Treaty, causing international friction.[71]

Harvard astronomer Jonathan McDowell found China's
maneuver disturbing because it marked "the first real escalation in
the weaponization of space [we had] seen in 20 years... [ending]
a long period of restraint."[72] Considering our military's need to
respond to threats from adversaries across (or orbiting) the globe,
it stands to reason that our government would see the need to react
by placing equipment in all compromising territories. Furthermore,
in reply to China's antisatellite test, Council on Foreign Relations
backgrounder Carin Zissis stated that through this action "China
shows off its growing military might in space to its neighbors and
the world.... From the U.S. perspective, China's capacity to destroy
satellites means it can target an American military weakness: the
reliance on satellites for intelligence gathering and the operations of
high-precision weaponry."[73]

With so much of our lifestyle, economy, and even military
defense system reliant on the function of the satellites within space,
many prominent figures state that it is high time we place a strategic, tactical presence within the same realm. The comparison

has been made between current conditions in space and "the Wild West, with a wide-ranging mix of government and commercial satellites, all of them sitting ducks."[74]

Yet simultaneously many who openly admit that military defense in space is forthcoming claim that it is merely to establish a presence or obtain placement as a precautionary measure, not as an offensive maneuver. While it is granted that the goal for the space force would be realized through creating a sixth division of our current military—to serve alongside the army, air force, marines, navy, and coast guard—many of these same individuals agree that it "seems unlikely that the space force will be sending troops to space on a regular basis, if at all. Instead…a space force would be much more focused on imposing military influence on current space traffic, which is mostly unmanned spacecraft…and also consolidating the way items in space are used to guide and assist military operations on the surface of the planet."[75]

While it remains a fact that unguarded satellites would leave open vulnerabilities for any correlating nation, the depth to which military tactics are currently being developed remains to be determined.

Astrophysicist, director of Hayden Planetarium in New York, and scientific communicator Neil deGrasse Tyson pointed out that opening the gateway to space could equal a new plain on which warfare could be waged. He asserts that although a treaty exists and new ones will be initiated to regulate space, there is always the possibility that those looking for friction will find ways to usurp the legislation's intent, or worse, these individuals may simply choose to violate the agreements. Certainly there is technology in place to monitor *what* is being launched and *by whom*, but this doesn't regulate intent. For example, the term *defensive* can be subjective. As the current Outer Space Treaty does not allow for *offensive* weapons in space, what equipment is considered offensive and defensive could depend on point of view and circumstance.

DeGrasse Tyson elaborates: "If I put a laser in space to disable something that might try to disable me, it counts as defensive, but it's really preemptively striking."[76] He furthermore explains that while it is necessary for a space force to develop a reliable asteroid-deflection system, such technology could be vastly misused, to our own demise. He explains in his book *Accessory to War: The Unspoken Alliance Between Astrophysics and the Military* that if technology can be used to deviate an asteroid from its path and away from Earth, then surely that same power could be used to channel something *toward* our planet in a devastating new form of warfare, and such abuse of the capability could be difficult to predict or prevent.[77] However, it should be clarified that the man is not opposed to developing a space force; in fact he advocates for the monitoring of NEOs and space debris, explaining that if the dinosaurs had had their own space force, they would still be on the earth. He likewise points out his pessimism toward man-made promises of cosmic amicability, stating that if people were capable of keeping the peace in space, Earth would likewise be a combat-free zone.[78]

And yet it remains that as long as others are routing technology, information, satellites, commerce, or even weapons toward space, those who do not keep up could find themselves compromised and vulnerable. Mark A. Bucknam is a professor of national security strategy and a course director at the National War College, a former air force colonel of thirty years, and the Pentagon director for Campaign and Contingency Plans for almost four years.[79] Defender Publishing was fortunate enough to send an agent to conduct a private interview with this extremely qualified expert, in which Bucknam said that many matters of national security are made up of a "lot of cross-over with Science and Technology." He explained that it was in 2007, when he was "participating in a joint effort with the Johns Hopkins Propulsion Lab,"[80] that the concern of a threat caused by a NEO or PHO was brought to his attention during a course on national security technology.

In his work titled "Asteroid Threat? The Problem of Planetary Defence," Bucknam acknowledges that it is not a matter of *if* but *when* the earth will again be struck by a large celestial body. Will that be biblical Wormwood? When a NEO or PHO of planet-threatening size does head our way, Bucknam stated that "nuclear detonations offer the only feasible hope of imparting enough energy to deflect the largest of PHOs,"[81] and yet Bucknam acknowledged in his interview with me (for this work) that NASA's studies have cast doubt on this particular course of action.[82] Perhaps NASA has confirmed in further research that this method of attack would be less effective than originally thought, or maybe it is because such weapons are banned by the Outer Space Treaty. To make an exception would require some sort of international collaboration to allow interference when an earth-shattering NEO is concerned—and yet it could open up its own new proverbial can of worms. Issues that would need to be ironed out before taking such measures would include the number of such weapons that would be allowed in space, the location and duration of placement, what would be done with such equipment once its service term is completed, and precisely who would monitor, maintain, and control them.[83] One wonders what type of international committee or dominating power would govern such a unified endeavor. (When recalling the implications of a one-world government found in Revelation 13, it makes a person wonder what role this space activity will play in prophetic manifestation. Also, what of the "power of the air" of the Book of Ephesians? But I digress...)

When asked if a comet- or asteroid-related electromagnetic pulse (EMP) were a possibility, he explained that this was not specifically a concern but added that "if you wanted to use a nuke to intercept and destroy an asteroid or comet, you'd want to intercept it far,

far from Earth—not within 300–400 miles up, where EMP effects would be generated. The further away from Earth you go, the less concern there'd be of any EMP effect from nuclear detonation."[84]

As for the space force currently being assembled, Bucknam was asked if, in his professional opinion, the organization was being amassed in response to the NEO/PHO threat, or if he believed it to hold more of an international property, such as militaries competing to gain weaponry advantages in space. His answer held that "the Space Force isn't being created to deal with NEO or PHO threats. It's due to military threats from countries like Russia and China." He adds that due to the current intentions of China and Russia, he does not believe that the space force will see asteroid threat mitigation as a priority.[85]

Despite how Bucknam believes the space force will prioritize the issue of asteroid threat, he maintains that cosmic threat matters are "a small probability but...[impact would have] extreme consequences." Where NASA's projection of Apophis' trajectory is calculated, Bucknam stated, "I'm sure they have a lot more data than they did originally....Orbital mechanics are sufficiently understood—I have confidence that they know what they are saying."[86] However, regarding the threats of comets versus asteroids, Bucknam emphasized one final point:

> I really want to be clear on the difference between asteroids and comets...most comets are only kilometers across. People need to understand that most NEOs are asteroids and they are predictable with certain speeds and orbits in or very near the same plane as the earth's orbit around the sun. Comets travel at much faster speeds and can come from any direction, even perpendicular to the earth's orbital plane, and they are unpredictable so we would likely have less warning time with a comet than we would with a large asteroid.[87]

FIRE FROM THE SKY?

On October 8, 1871, an eruption of flames throughout the United States' Midwest region seemed to materialize. Three states—Michigan, Wisconsin, and Illinois—were impacted almost simultaneously by the outburst of these three separate wildfires, with flames killing thousands of people and destroying millions of acres throughout the cities and surrounding regions of Peshtigo, Holland, Manistee, Port Huron, and Chicago.[88] Over the years, many have pinned the blame for the Great Chicago Fire on Mrs. O'Leary's cow, which, it has been said, kicked over a lantern in the O'Leary barn, igniting the infamous fire, which "destroyed thousands of buildings, killed an estimated 300 people and caused an estimated $200 million in damages."[89] However, a deeper look at this outbreak renders a different, more cosmic possibility.

Comet Biela, originally discovered in 1821, was on an orbit that, at that time, cycled near earth "every six years and nine months. It was last seen in 1866 and never showed up in 1872."[90] Robert Wood, retired physicist of McDonnell Douglas Corp., asserts the theory that combustive materials that fell to Earth as a result of the comet breaking up were the culprits of these almost simultaneous fiery outbreaks. According to Wood's theory, Biela's comet had orbited too close to Jupiter around 1845, causing it to break into "two large fragments," after which "astronomers noted a 1.5-million mile, 15-day gap between the two pieces." The aftermath of this close brush with Jupiter, according to Wood, would have placed one of the fragments within close proximity of Jupiter, whose gravity would have altered the speed and trajectory of the pieces, pointing the smaller at Earth, which was subsequently impacted in 1871.[91]

The sheer coincidence of these fires hitting all within hours of one another makes it believable that more was at work here than a mischievous, lantern-kicking cow. However, beyond the quirky happenstance of the timing of the three fires were the ways in which they were destructive and the unique aftermath that became evident when the blazes had settled. One unusual result related to these incidents was the fact that many of those who perished showed no evidence of burns, which Wood stated was more "consistent with...the absence of oxygen or the presence of carbon monoxide above lethal levels," which would have been more easily explained by the descending "fire balloons"[92] spoken of by surviving eyewitnesses than by a regular forest fire. Likewise, observers later reported witnessing what they could only explain as spontaneous combustion sparking the ignitions.[93]

Within this twenty-four-hour period "an area of land the size of Connecticut was burned"[94] between these three coincidentally originating fires. For three blazes of this magnitude to occur within hours of one another in proximal areas, one is tempted to believe that the theory of Biela's comet being to blame holds water. To further the argument, in 1990 a work crew drilling to install a water pipeline beneath Lake Huron discovered a huge impact crater two hundred feet below the lake. Rocks from this source were analyzed and found to be either from a meteorite or of volcanic nature, fortifying the probability that "Michigan...parts of Canada, [and] Illinois are ground zero for an active meteor strike zone."[95]

This theory has been received by many as highly controversial and has spurred much heated debate. Additional fires, impacting nearly ten other states and speckled throughout Canada on the same day, have been added to this theory by some, which are equally met with vehement denial. Strangely it is also stated that two decades after the fires, Michael Ahern openly admitted to concocting the story about the O'Learys' cow, vindicating the creature from blame for the disaster.[96] Despite this change of tune, the poor animal remains the object of many cartoons, children's stories, and other pop culture media, pointing the finger at it as the cause of the Great Chicago Fire.

The idea that these fires were started by a colliding comet bears merit, and in many ways better explains how such suspiciously similar events could have occurred in different places over a series of hours. Additionally, if this theory is correct, then the event could confirm the very theory that some have worried over regarding Apophis: that a NEO traveling peacefully within proximity to Earth could interact with some other body in space, altering its course, and then hurl toward our planet.

Considering the monitoring of satellites, defensive weapons, increasing commercial industry, and even asteroid mitigation, one would presume there are many reasons pertaining to national security to launch a space force. However, for those who wonder if there is more to the story, there are the additional mysteries of space that provide a breeding ground for speculation, especially in light of the vagueness that current descriptions and explanations of the space force's projected functions offer. When the reader begins to explore such possibilities, he or she may begin to wonder if there is validity to the concept that there is more at work in this endeavor than meets the eye. Are we looking for colonization in space or even more?

THE ENIGMATIC 67P/CHURYUMOV-GERASIMENKO

When the European Space Agency's Rosetta Mission landed the probe Philae on what was believed to be a comet, it was a successful day for space researchers around the world. After all such a feat had never been accomplished before.[97] Immediately conspiracy theories and speculation began, and many questions have remained unanswered. When it was revealed that the comet was emitting a strange sound that many can only describe as a song, theories emerged from all over the globe. Some made attempts at natural explanations for the phenomenon, such as vibrations "set off by a stream of charged particles ejected from the surface of the space rock."[98] Others stated that the mass was not a comet at all but an extraterrestrial mystery craft that had previously communicated with NASA and that the entire endeavor had been an attempt to communicate with aliens but *disguised for purposes of publicity* as a comet-exploration mission. Furthermore, the nature of the "song" itself has been the subject of much speculation, some saying that the message is merely a greeting to life throughout the universe from beings looking to make contact, and others perceiving it as a warning from a race with superior technology.[99]

I recall that at the time this was all in the news, I reached out to my good friend Dr. Chuck Missler to ask if he knew anybody in a research or intelligence agency who might have access to a supercomputer or other system that could slow down the sound waves on the Rosetta comet to check for any repeatable patterns. It was a long shot, but I wondered if there might be some type of replicating synthetic (nonorganic) code in this sound that might indicate it was not just wind sounds passing the object. If so, it would be an astonishing discovery. Chuck knew a man who had been involved very early in developing audio technology whom he said was a bona fide genius. I cannot recall the fellow's name now, but I sent him the audio recording from the comet without telling him where the sound came from. The only response I received from him was his asking if anybody else became sick from listening to the audio recording. When I didn't hear more, I reached out to two friends (whose names I must protect) who are special technicians for space missions with top secret security clearance with NASA and took a stab at my request using Fourier-based deconvolution for resolving oscillographic signals. They sent me the following short report:

Comet 67P "Song" Captured by Rosetta Spacecraft
December 2014

• The European Space Agency (ESA) has released an audio track which mimics a "song" coming from comet 67P.

• The "song" has a measured frequency spectrum falling between 40–50 millihertz, which is far below the normal hearing range of 20 Hz to 20 kHz. In order to hear anything at all, the ESA investigators had to shift the frequency spectrum to the human audio range, by raising the frequency spectrum by a factor of 10,000–100,000. It appears that this was a scalar multiply operation.

• According to ESA, the "song" is produced by temporal oscillations in the magnetic field surrounding the comet as it moves through space. However, exactly how the sound is produced in this fashion remains a mystery.

• And this is how the "song" was recorded—via an on-board magnetometer on the Rosetta spacecraft. The magnetometer was designed to "monitor how the comet would interact with the solar wind and with particle plasma emitted by the sun." The "song" was picked up at approximately 60+ kilometers away from the comet and continued uninterrupted as the spacecraft moved closer.

• Some thoughts from us both about the "song," without the benefit of a full frequency content analysis

Obviously, a narrow audio frequency band between 40–50 millihertz is at the heart of the transmission. This has the advantage of better communication over long distances.

The "song"/signal has every appearance of being modulated in frequency and tone, which mimics how ordinary communication with data bits is designed.

One reads about the song being the effective result of random or manifold complex magnetic field interactions, *but the signal indicates something more than noise in a pure random process. The speculation of magnetic field interaction producing this type of signal modulation defies ordinary experiences of physics* [emphasis added].

It is conceivable that the comet is oscillating itself via some complex resonance modes.

Friend number one comments:

On a side note, I think it is very interesting to consider the use of a comet as a vehicle for a "message in a bottle." It's a great application, if there are other civilizations out there, then it would make sense to assume that they would explore the use of comets as a first step to deeper space exploration. And given the size of the universe, you can't send a signal out to all of space (that requiring energy levels approaching the infinite). So the strategy would be to put something on a comet like a timing circuit that would start counting with time zero being when the comet passed closest to earth. That way if someone found the comet and the radio tower signal, they can measure and correlate the time and get a good estimate of solar system that was at that location at time zero, based on the comet trajectory. That way they could send a strong signal back in the direction of earth instead of 4-pi steradians. Of course this and/or all other comets we see may just be stuck around our solar system, so not sure that concept holds water. This makes more sense if the comet is more like an exploration vehicle traveling through the universe.

Friend number two comments:

The message in the bottle has the capacity of indicating two things: real signal content, and a stellar geography to potentially extrapolate the source position.

What is really required is a more thorough mathematical analysis which can look at the song fluctuations in both frequency and tone level around the modulating base. Obtaining the original audio spectrum from ESA seems impossible. However, we will try and take what is there and work along these lines.[100]

Another item related to the 67P that has stumped experts is the fact that the surface seems to hold an icy dust that contains organic molecules, which experts say likely make up half the body's constitution, meaning it "belongs to the most carbon-rich bodies we know in the solar system."[101] In addition to the strange dust and mysterious sound emitting from the celestial anomaly is the shape of the body itself, which some have compared to an alien bust.[102] Some assert that the shape could not be natural and even claim that the mass is in actuality an ancient alien spaceship.[103] On top of allegations such as these, photographs taken of the comet show mystery markings that some claim are UFOs hovering over the surface of the mass, along with another mysterious silhouette that some state is a radio transmission tower.[104] Other, less conspiratorially minded individuals claim that the shapes in such images are mere craters and fissures in the surface of the body.

Because there is so much still to discover about the mysterious realm beyond our own planet, and since there is so little literal communication regarding the strategical functions that are planned for once we make our presence known in space, speculation still presides in the minds of many as to what will really be found or accomplished once that presence has been established. The possibilities seem endless between commercial development, military maneuvering, colonization, or even the search for extraterrestrial existence, which will continue to, if nothing else ever comes of it, propel man's curiosity and compulsion to seek and conquer all territories he can possibly reach.

And yet one wonders if in this pursuit we cross a detrimental line.

ARE WE GOING TOO FAR?

By now it should be apparent that there are many reasons currently speculated regarding the true motivation behind the need for a space force. And to be sure, many of the more practical reasons, such as satellite and military defense, shed doubt on the reality of the need for asteroid mitigation preparedness. However, the prophecies found in Revelation 8 that discuss the very *real* disclosure that there *will be* objects colliding with Earth show that these will be instruments of judgment.

With that in mind, one must consider the question "What type of event will kick-start this onslaught of wrath from on high?" Could it be, perhaps, that our own tampering in space will be what launches the spiral of events accompanying the seven trumpets? This may seem like a fantastical concept, but imagine the possibility of warfare in space, asteroid mining, or even colonization. Consider all the times in the history of mankind's innovations when equipment failed, operators made an error, or even the wreckage of warfare left an earthly zone destitute or rampant with devastating consequences. Certainly in space if we can deflect an asteroid on purpose, then with enough heavy activity within that realm doesn't it seem possible that we could somehow knock a cosmic body off its trajectory, causing disastrous consequences that backfire?

Perhaps it will be mankind that initiates the very events that bring about their own judgment...

In Genesis chapter 11 a story is told of a time when the people of earth spoke in one universal language. This unified collective came together and made a plan to "build…a city and a tower, whose top may reach unto heaven" (Gen. 11:4), which God subsequently inspected. The Lord's response to this tower was: "Behold, the people is one, and they have all one language; and this they begin to do: and now nothing will be restrained from them, which they have imagined to do" (v. 6). Because God did not approve of the people's undertaking, He scattered them throughout the earth and confused their language. Consider how the *Lexham Glossary of Theology* portrays the attitude of the population at Babel toward the almighty God:

> In Gen 11:1–9, the people of Babel attempt to build a tower to reach the heavens, where they believed the gods dwelt. The tower was probably a Mesopotamian ziggurat, a large step pyramid with a temple at the top. The tower of Babel is a prominent biblical symbol of human arrogance in seeking equality with him, autonomy from him, and/or to reach him on their terms rather than his.[105]

The concept that God would be put off by mankind's desire to build a tall tower confuses many people. After all, was it *really physically possible* that these individuals would have built a structure so tall that it ascended beyond our planet and into another world? The idea seems far-fetched, and yet even God Himself recognized that "nothing…[would] be restrained from them, which they have imagined to do" (Gen. 11:6).

In order to understand the great sin involved in this undertaking, one must take a closer look at the psychology of the day. Titus Flavius Josephus wrote a bit of this, explaining that these ancients were well aware that in their semi-recent history an overpowering and devastating flood had wiped out almost all life on the planet. There was no argument regarding whether this deluge had taken place, but despite God's covenant, as to whether He would destroy the earth through flood again (Gen. 9:11–17), there was anxiety among the people that a flood could reoccur.[106]

In response to this vulnerability it was the people's reaction to build high structures. While ziggurats held other significances to these ancients, their *height* was designed in an attempt to take matters into their own hands and escape judgment. Beyond this some extrabiblical sources claim that this tower featured an "astral temple at the top."[107]

On this matter Josephus wrote:

> Now it was Nimrod who excited them to such an affront and contempt of God. He was the Grandson of Ham, the son of Noah, a bold man, and of great strength of hand. He persuaded them not to ascribe it to God, as if it was through his [God's] means that they were happy, but to believe that it was their own courage which procured that happiness. He also gradually changed the government into tyranny, seeing no other way of turning men from the fear of God, but to bring them into a constant dependence upon his [Nimrod's] power. He also said, "He would be revenged on God, if he should have a mind to drown the world again; for that he would build a tower too high for the waters to be able to reach; and that he would avenge himself on God for destroying their forefathers!"[108]

This insight reveals a new level of rebellion toward God, carried out through the act of building the tower. Its very construction sent one single statement to heaven: *People do what they want, when they want to do it, and on their own terms. We do not submit to Your authority, and if You try to flood us (or otherwise send judgment upon us again), we'll be ready.*

In the beginning God created a setting wherein man was given permission to dominate. Consider the following passages in Genesis regarding the jurisdiction allotted to man:

And God said, Let us make man in our image, after our likeness: and let them have dominion over the fish of the sea, and
over the fowl of the air, and over the cattle, and over all the
earth, and over every creeping thing that creepeth upon the
earth. So God created man in his own image, in the image of
God created he him; male and female created he them. And
God blessed them, and God said unto them, Be fruitful, and
multiply, and replenish the earth, and subdue it: and have
dominion over the fish of the sea, and over the fowl of the air,
and over every living thing that moveth upon the earth.

—Genesis 1:26–28

And the Lord God took the man, and put him into the garden
of Eden to dress it and to keep it.... And out of the ground the
Lord God formed every beast of the field, and every fowl of
the air; and brought them unto Adam to see what he would
call them: and whatsoever Adam called every living creature,
that was the name thereof.

—Genesis 2:15, 19

Now more than any time since the days of the Tower of Babel
the people of earth show similar parallels to that moment in time.
Thanks to the digital interface, the world's populace is continually
interconnected, and while we do not have the same cultural back-
grounds or speak the same ethnic language, we are all capable of
communicating and acting in a collective manner. Now, as we begin
once again to find a way to "reach unto heaven," as was attempted
in those days, are we inviting judgment upon ourselves for leaving
the region over which God gave us dominion?

The Bible never says that God gave us any rights to space. In fact
He specified the zones over which man was given dominion: the
sea, the air (where birds fly), the land (cattle), and *over all the earth*.
God created the firmament (Gen. 1:6–7), and the stars, moon, and
sun (Gen. 1:14–18). When man tried to leave that zone by building

the tower, God stopped him. Scripture does not tell us that He gave us the authority to dominate more than the earth itself. In fact God says, "Though thou set thy nest among the stars, thence will I bring thee down, saith the LORD" (Obad. 4). Perhaps we are forgetting our role and are taking things a bit too far. Will this usurpation be part of what causes judgment to fall on us?

Chapter 4

PLANETARY PING-PONG

I N THE SECOND chapter I asked readers to imagine seeing a horned, fiery serpent, possibly as large as three miles wide, plunging toward the earth at an immeasurable speed. This terrifying monster seems to swim across the sky, past the stars, descending closer to the earth until, making contact, it plunges into the ocean, its massive form sending a sequence of tsunamis measuring six hundred feet apiece in height slamming into the coastal terrain of regions across nearly half the world, infusing the atmosphere with scorched particles of aerosol and vapor. The resulting blistering culmination of moisture and extreme heat in the earth's atmosphere subsequently combusts into a series of high-velocity hurricanes, which turn their deadly gaze upon the yet unaffected hemisphere of the world. So much debris is released as a by-product of the initial impact and consequent devastation that for about a week darkness covers the sky worldwide as the entire landscape is pounded by hurricanes and similar atmospheric annihilation. By the time the waters finally settle, the storms subside, and the sky grows clear, most of the life on Earth is dead. According to Los Alamos National Laboratory archaeologist Bruce Masse, this is precisely what those alive on the earth in the times of Noah saw, and *this* is how the deluge of that time took place.[1]

In addition to scientific evidence used by Masse to fortify his case, the archaeologist interprets an ancient petroglyph etched into place by a Native American shaman as an additional reinforcement of this detail. This etching displays a horned, serpentine image moving among the stars, heading for our planet. Masse explains that the lore of cultures all across the globe tells of a planet-devastating comet striking Earth and causing a flood that wiped out most of the life on our planet. He believes that this comet struck the Indian Ocean on May 10, 2807 BC, after coursing around the sun—a passage that exposes a comet to solar winds that shift the orbital body temporarily, whipping its tail around to precede the main body for a time, creating the illusion of a horn on its head.[2]

Cultures around the world—Hebrew, Mesopotamian, Hindu, Chinese, indigenous South American, Indian, Native American, and many others—are the source of the 175 flood accounts that Masse examined.[3] His findings were fascinating: many describe such a cataclysmic event, followed by a flood that wiped out nearly all life on earth. Similarities among the tales included the fact that fourteen of them referenced a solar eclipse, half mentioned a torrential downpour, and one-third recalled tsunami-type waves. "Worldwide they describe hurricane force winds and darkness during the storm. All of these could come from a mega-tsunami."[4]

Masse understood that mythology offered interesting theories, but he wondered, Did the physical evidence add up?[5]

Joining forces with University of Wollongong geomorphologist Ted Bryant and Lamont-Doherty Earth Observatory at Columbia University professor Dallas Abbot, Masse and these two qualified individuals founded the Holocene Impact Working Group, an agency that continued the search for evidence of such a mega-tsunami. This crew analyzed the type of coastal damage and debris left behind by cataclysmic oceanic asteroid collision: if it had happened as they believed, they expected to find enormous chevrons wedged into the coastline near the suspected impact area, embedded with "deep-ocean microfossils" that would have been stirred from the deep by the tempestuous waves. As expected, these very impressions were found in the anticipated coastal regions, made up of material from the ocean's floor[6]—so much debris from the deep in fact that "each [chevron] covers twice the area of Manhattan with sediment as deep as the Chrysler Building is high." Furthermore, the residue amid the fossilized material includes a blend of metals characteristic of asteroid or comet impact.[7] On top of this startling find, similar structures four miles from the water's edge at the suspected point of impact indicated waves traveled twenty-five miles along the shoreline, a phenomenon these experts say could not have been explained by any element other than oceanic asteroid collision.[8] All of this evidence fortifies the case that the eighteen-mile-diameter crater recently discovered nearly thirteen thousand feet beneath the surface of the Indian Ocean is the location of a cataclysmic cosmic impact such as the one described here. While some argue against the concept of a cosmic event, stating that these chevrons have been formed by millions of years of erosion, the experts at the Holocene Impact Working Group remain convinced, asserting that other elements do not support this position. For one thing, the geographical spread of debris from the ocean's floor was in some places "deposited along whole coastlines," indicating waves larger than any tsunami the modern world has seen.[9] Additionally mere natural elements such as erosion or volcanoes do not explain the existence of the cosmic metallic blend found within the fossilized residue along the chevrons. Furthermore, separate craters analyzed and dated to a different period than that suspected to have initiated the flood have rendered the fascinating presence of diatoms—"microscopic sea organisms"[10]—found fused together with tektites, a molten, "glassy substance formed by meteors."[11]

As these and other experts continue to glean valuable information about previous cataclysmic cosmic events experienced by our planet, some who see a biblical parallel begin to ask the question "Could this happen again?"

"But as the days of Noah were, so shall also the coming of the Son of man be" (Matt. 24:37).

EARTH TO REENTER THE METEOR STORM THAT BROUGHT ABOUT THE TUNGUSKA EVENT

In a previous chapter we discussed the Tunguska event of 1908, wherein an entire forest was destroyed when what many believe was a meteorite struck in Siberia, destroying virtually all life in a region covering eight hundred square miles. The Taurid swarm is thought to be a large contributing factor to the circumstances that brought about this event. It is a conglomeration of meteors made up of the debris of comet Encke, which the earth passes, each time at varying proximity. Usually rubble that enters Earth's atmosphere burns up, leaving our planet unscathed and having the added benefit of viewing a meteor shower. However, this swarm increases the odds of a large impact, according to Western Ontario University researchers, and is likewise considered "one of the three space phenomenon that could result in a catastrophic collision." Since it is projected that Earth will pass "within 30,000,000 km of the center of the Taurid swarm,"[12] researchers will be watching closely in an effort to learn if our planet is at risk of a repeat Tunguska event, and if so, *when.*

IF A COSMIC EVENT LAUNCHED THE FLOOD OF NOAH...

The Holocene Impact Working Group members are not the only authorities who believe the many instances of cultural lore referencing a worldwide flood event that took place as a result of cosmic collision. Donald W. Patten's enlightening book *The Biblical Flood and the Ice Epoch* breaks down in scientific detail how astral activity likely caused the flood of Noah. The work is so extensive that it cannot be adequately covered in this book, but a few details enhance our study, so we will briefly touch on them. The first point to be made is that while Earth has been battered sporadically by objects from space throughout its lifetime, the events launching the flood were by far the most devastating. He explains that through a cosmic event the tides of the earth surged, that fluid magma in the form of lava escaped the earth's crust, "the fountains of the great deep [were] broken up" (Gen. 7:11). As a result the waters rose, or "increased...upon the earth" (Gen. 7:18), and the ensuing residue released into the atmosphere caused electrical reactions within the earth's atmospheric system, which caused torrential rains and other devastatingly erratic weather for a subsequent amount of time. Patten outlines that a "large astral body, travelling in an eccentric (highly elliptical) orbit, would be sufficient to cause a great flood through tidal mechanisms."[13] In his argument, which he acknowledges has elements of history, geology, and astronomy, he explains that unusual elements—such as marine fossils uncovered at extraordinary elevations in the Swiss Alps and terrestrial fossils found low in the ocean—confirm the occurrence of a great flood in earth's past.[14] As for the specifics of *how* a cosmic collision initiated the flood of Noah, his theories are very similar to that of the Holocene Impact Working Group. In a nutshell, after a devastating cosmic event (likely an impact with an extremely large comet) that left interference between the earth and our moon, causing a "major conflict of gravity within the Earth-Moon system,"[15] unprecedented tides rose *simultaneously* with torrential rain. This concept is confirmed in Genesis 7:11: "The same day were all the fountains of the great deep broken up, *and* the windows of heaven were opened" (emphasis added).

An additional factor Patten cites is the heat exchange between large amounts of astral ice smashing into the ocean (likely -200 degrees, and falling "at a rate of several hundreds of feet per hour"[16]) and the magma released from the earth's crust during this earth-shattering smashup. Essentially, as this heated substance escaped Earth's fractured core, it expanded against the temperature-contrasted icy water above, causing the volume (and volatility) of the substances to raise. (If you have ever heated an empty pan and thrown a teaspoon of cold water onto its surface, you have seen a tiny representation of this concept.) Added to this explosiveness beneath the water's surface is the irreconcilable series of mega-tsunamis caused by the impact itself. Each separate condition (the fire-ice smash below the water and the colossal waves above) subsequently perpetuated the other, friction causing plumes of heated gases from the earth's core to boil through the water and into the atmosphere, contributing to hurricane activity, which was accompanied by a torrential downpour.

Patten explains that the reason excessive numbers of fossils can be found misplaced at high elevations is the waves that occurred during this time. As a result of the extreme weight of water that rose so quickly, living creatures were compressed beneath the volume and fossilized immediately, having no time to decay. Sweeping tsunamis, rolling so fiercely from the ocean's floor as a result of the magma below, then picked these items up and sprayed them across entire mountain ranges.[17] In addition to causing such tidal frenzy, the fluid magma beneath the water, when it erupted, released a pressurized explosion that, according to Patten, altered the terrain of the land and caused entirely new mountain ranges to form as a by-product of the centrifugal thrust.[18] Even after the rain stopped, tides continued to rise until all land upon Earth was underwater and most of the life on Earth perished.

SCARS IN SPACE

Another indication that Earth has at some point been pummeled by a large cosmic body can be found in the crust of the earth. The K-T boundary is defined as the layer within the earth's crust that separates the Cretaceous period from the Tertiary, or, in simpler terms, the separation of the dinosaur era from the non-dinosaur era. This layer is found across the earth, deep below the surface of each continent, and contains extremely high levels (between 30 and 130 times the anticipated amount) of iridium, an element rare on earth but plentiful in asteroids. In fact the level of iridium in the K-T boundary, when tested by physicist Luis Alvarez and geologist Walter Alvarez, was found to match the level of the same element in meteorites. Experts estimate that the meteorite that must have struck Earth in order to create this layer within the earth's crust was likely ten kilometers in diameter and would have detonated with the force of "100 trillion tons of TNT," and it is suspected that this disastrous event would have involved a dust covering (similar to that of the Shoemaker-Levy—more on this later), which left the earth dark for a period of years. Theories then state that vegetation suffered, eventually starving out habitants of the planet, such as dinosaurs.[19]

While some logistics of this theory are debated—such as precisely how many years ago this occurred and exactly which asteroid did the job—the layer itself provides evidence that the earth has at some point been victimized by a devastating cosmic-planetary event. In fact it is no secret that the earth has been impacted many times, despite the fact that the evidence hides in plain sight. In a report released by the Directorate of Strategic Planning at the headquarters of the US Air Force, Lindley Johnson stated that "over 160 impact craters have been identified on Earth and more are discovered all the time. Earth has been hit every bit as often as the Moon, but because Earth is a living planet with large ocean areas, weather and hydrologic cycles, and moving tectonic plates, impact formations get eroded or covered up."[20]

However, Earth is not the only body in our solar system that shows signs of large-scale astral wreckage. Much of our solar system bears the evidence of a "planetary ping-pong" session that has taken place at some point in its history, although the date has been the subject of debate. This excerpt from *Unearthing the Lost World of the Cloudeaters* elaborates:

> During this event, the solar system was pummeled with enormous meteorites over a period of time. Earth, Mercury, Venus, and Mars were likely the survivors most heavily pounded by the crossfire…the earth's previously vertical axis shifted. Uranus was whacked onto its side during the event, and Venus' orbit was completely reversed. Planetary debris clustered into bands and scattered throughout our solar system, forming trajectories of water, dust, and debris, which we now call "comets." It is even surmised that Caribbean Sea is an impact crater from this event.[21]

In addition to these scars in space is the possibility that Mars at one time had an atmosphere capable of sustaining life. It would appear that this planet held water and could have even supported vegetation, before an onslaught of meteorites the size of Manhattan destroyed this asset. Furthermore, stress ridges and impact craters on the surface of Mars show that it has indeed taken a planetary beating. Neptune and Pluto both show interruption in their planetary orbits and lunar patterns, and multiple planets in our solar system acquired new rings, courtesy of the floating space debris that was unleashed by the event.[22] Our own moon was recently discovered to be in much worse condition than scientists once believed, with recent analysis revealing that asteroid impacts and craters have been discovered to run as deep into the moon's surface as twelve miles. It is now thought that "ancient impacts could have substantially fractured the lunar surface,"[23] weakening the moon's overall solidity. Uranus is a unique planet from the others, as it rotates sideways and seems eerily isolated in its frozen habitat. Durham University scientists decided to run some simulations in an attempt to understand these oddities with better clarity and found that a likely possibility was that at some point in the planet's past it had collided with a cosmic body twice the size of our planet, which would have sent the larger body cockeyed and is likely the source of the dust, debris, and even moons that accompany Uranus in its orbit.[24]

The list of scars in space is endless and could be (and *is*) the subject of many books. Suffice to say, each planet bears the mutilations of our solar system's bumpy history. However, despite theories that Earth has had at least one cosmic event in its past that wiped out nearly all life, it would be fair to say that it did not take the worst hit in our solar system.

After all, Earth is still here.

Bode's law

Bode's law is an empirical rule by which some experts have used mathematics to predict the spacing between each planet in our solar system and our sun.[25] Utilizing this formula, authorities found a shocking surprise: there is a fifth revolution niche that is mysteriously vacant, exhibiting in place of its once prominent occupant only a solitary memorial: an asteroid belt approximately five hundred miles in width, "its girth clustered with planetismal-sized chunks, its culminations diminishing down to fine dust."[26]

In simpler words, there is a five-hundred-mile mass of rubble occupying an otherwise empty planetary orbital track in our solar system—evidence of a missing, *destroyed* planet.

Missing planet

The body would have orbited between Mars and Jupiter, likely would have been visible within the earth's sky both day and night, and is theorized to have been a water-bearing planet, based on the icy debris its remains have lent to the rings of neighboring planets.[27] Scientists at the Southwest Research Institute in San Antonio ran thousands of simulations in an attempt to determine if the rubble was a missing planet, and they came to the conclusion that it was conceivable not only that a large planet once occupied this orbit but also that it was unlikely that our solar system could have reached its current placement *without* the dynamic of a fifth large planet (in addition to our known four: Uranus, Neptune, Jupiter, and Saturn). Beyond this the research institute quoted David Nesvorny's assessment: "Just as an expert chess player sacrifices a piece to protect the queen, the solar system may have given up a giant planet and spared Earth."[28]

Some say that this crazy round of planetary ping-pong took place during the Late Heavy Bombardment period, which took place approximately four billion years ago,[29] while others say that our solar system is much younger than this and the damage has gradually stacked over the centuries. Additional individuals claim that around 3,000 BC increased record of astronomical occurrences suggest elevated activity within our solar system, setting the scene for cosmic activity that may have brought about the flood of Noah, while another group asserts that the missing planet was a place called Rahab, referred to in Psalm 89:10, which says, "Thou hast broken Rahab in pieces," claiming that this was the world of the fallen angels, which God destroyed in judgment.

Regardless of where an individual stands among these debated points, the evidence of damage within the solar system remains. One can decide what one believes to be the cause of such a round of planetary ping-pong, but there is *no doubt* that such events occurred. With this in mind the sheer brutality of the bashing that must have occurred as these cosmic bodies thwacked about in space becomes unfathomable. Perhaps it was as simple as a large body's course somehow being altered, causing multiple celestial masses to derail and collide, like one giant cosmic train wreck.

Or perhaps it was more.

Perhaps God used a cosmic event to incite His judgment on a fallen planet, after His select few entered an ark. We know from the Book of Revelation that He intends to judge the earth with astral catastrophe in the future, and evidence seems to suggest that this could have been the way God dispersed His wrath in the times of Noah.

With this in mind one must wonder, if the coming of the Son of Man will be as it was in the times of Noah (Matt. 24:37) and judgment from on high was dispersed via astral catastrophe in Noah's day, then it's very possible that Christ's return will be foreshadowed by similar cosmic events. The idea brings new light to the biblical description of the day of the Lord:

> But the day of the Lord will come as a thief in the night; in the which the heavens shall pass away with a great noise, and the elements shall melt with fervent heat, the earth also and the works that are therein shall be burned up.
>
> —2 PETER 3:10

SHOEMAKER-LEVY 9

It is interesting to note that in the widespread ancient lore refer-encing an ancient flood, some cultures made note of total darkness coming over the earth for a time. Could such an event ever happen to Earth? Surely such things are referenced in prophetic scripture (more on this later), but what type of cosmic disaster would it take to make the earth go dark, even during the day?

On March 18, 1993, Gene and Carolyn Shoemaker, along with David Levy, discovered the aptly named comet Shoemaker-Levy 9, which was subsequently destroyed in a collision with Jupiter in 1994. The astral body is theorized to have been a single body (later estimated to be about 1.5–2 kilometers in width) that migrated too closely to Jupiter in July 1992 and, as a result of the strength of the large planet's gravity, was pulled in by its tidal forces and unable to orbit out of harm's way. By its discovery in 1993 the comet was already segmented into over twenty pieces, and it met its demise the following year. This unusual manifestation offered scientists and authorities a rare opportunity to observe an astral body as it orbited a larger one with strong gravitational forces and to even-tually witness its final moments. NASA's orbiter, the Galileo, was nearby to capture images of the entire event, while other equip-ment, such as the Hubble Space Telescope, the Ulysses, and the Voyager 2, was involved in the observation and study of this event, which took several days between July 16 and July 22 of 1994. When Shoemaker-Levy crashed into Jupiter, the "fragments smashed into Jupiter with the force of 300 million atomic bombs…[which] cre-ated huge plumes that were 2,000 to 3,000 kilometers…high, and heated the atmosphere to temperatures as hot as 30,000 to 40,000 degrees Celsius."[30] As a result Jupiter's main ring has some scar-ring, has become tilted by about 1.24 miles, and nearly twenty years later still shows signs of interference, as do Jupiter's atmospheric rhythms.[31]

When the collisions were over, a blackened array of scars loomed over Jupiter's cloud system, darkening the impacted area until subsequent winds over the planet dissipated the haze, taking more than a month.[32] After the fact, NASA stated that if a similar object were to impact Earth, the results would be disastrous. If a blackened conglomeration of astral matter similar to Shoemaker-Levy were to linger in Earth's sky, it would likely bring earthen temperatures down and leave our planet devoid of sunlight. Worse, NASA said that if this condition lasted long enough, vegetation could perish, with all other forms of life soon to follow.[33] Milton Kazmeyer of SeattlePi elaborates: "Upon impact, vaporized dirt and rock would fill the atmosphere, blocking sunlight and creating a state of near-permanent darkness and winter-like conditions," which would later be followed by severe heat surges as a result of an increased greenhouse effect once the initial dust cloud cover dissipated.[34] Of course the severity of this by-product would depend on the size of the cosmic force, but in consideration of the size of the mass suspected by some to have caused the flood, these concepts are feasible.

In addition to these potential dangers another risk factor remains: if an asteroid or comet were to significantly damage our ozone layer, the exposure of remaining life on this planet to the ultraviolet rays that did manage to reach the surface of the earth would be vastly increased and much more dangerous, increasing "global cancer rates, [affecting] the growth of plants and animals, and possibly [leading] to genetic mutations"[35] as time passes.

In light of these implications one begins to wonder if we on Earth can even begin to fathom the myriads of possible devastations that could be brought upon us if we were to experience an earth-shattering astral event. Is there any way to predict all potential emergencies and prepare for them? Even if it were in our power to anticipate and make provisions for the varying ways that we could be threatened by an astral event, could we do so in quantities large enough to take care of the entire region, an entire continent, or even the whole world?

With questions such as these in mind, a person must wonder if those in the know *really would* tell the general public if there were danger of Earth being struck by a NEO. After all, with such cataclysmic implications as discussed in this chapter and such feelings of inadequacy in the wake of such disaster, surely the general public would begin to panic. And with anticipated possible collision dates years into the future, there would be ample time for imaginations to run wild and paranoia to run rampant, tearing through communities and wiping out order as we know it. Perhaps the announcement that all is well and that we expect no threat from Apophis, 2018 LF16, or any other NEO will remain, if for no other reason than a means of crowd control.

Would "They" Tell the General Public?

What NASA is most worried about

In the simulation exercise conducted by NASA discussed earlier in this book, one aspect of preparedness that became a point of concern was anticipating and preparing for the needs of the populace of impacted and surrounding areas. During this exercise it was decided that any evacuation deemed necessary held dire consequences if not carried out carefully and strategically. Before movement of large crowds was carried out, a course of mobility needed to be outlined in detail and communicated to all authorities assisting with the endeavor. Authorities agreed that a mishandled evacuation was a greater risk than the actual meteorite impact. The need for one single authority—properly furnished with cutting-edge equipment for the purpose of graphic and topographic communication and with the ability to analyze population density geographically—to act as communicator with both media members *and* those who would oversee the relocation of people

whose location may be compromised by the disaster was immediately identified as crucial. Advance mapping of prime departure
routes along with preidentified optimal destination points would be
necessary for all potential impact locations as well. This authority
would need adequate personnel to handle the evacuation and destination coordination planning in addition to ample staff to handle
the many press and media inquiries the authority would certainly
be flooded with. Decisions would need to be made such as when
and how to disclose to the public that potential relocating would
be required—a premature announcement could cause mass panic,
resulting "in rash actions by others and the presence of alternate
'authorities.'...Suggesting a plan too early might create disruptive
movement and counterflow." Furthermore, an anticipated impact
location alters the potential course of such strategies, as a land strike
would be accompanied by an assessed fireball and blast radius, surrounded by further zones that would be susceptible to hurricane
weather and thus would need to be estimated and cleared. An oceanic impact would likely result in a tsunami response, requiring
a different evacuation strategy altogether. Furthermore, if the collision zone were near volcanic surfaces, the ensuing ash could
become dangerous for certain methods of transportation (such as
flight), while the risk of high winds increases the chances of elevated debris plumes. In a worst-case scenario such a disaster would
strike in an area where nuclear facilities are hosted, meaning not
only would the ground-level environment be contaminated by
these deadly materials, but debris plumes and winds could spread
these substances countless miles, making the sky lethal as well.[36]

Beyond attempting to evacuate the masses and bring them to
safety, authorities in such a situation would have the added consideration of meeting the physical needs of the throng. For example,
once the seemingly insurmountable job of relocating has been
completed, there are issues of food, water, shelter, clothing, medical

supplies, and even security. Certainly bringing individuals to a safe location would be only the beginning of this arduous task. Often such events disarm the power grid of the surrounding area, which would mean completing the evacuation and meeting immediate needs without the benefit of electricity or phone lines, and sometimes those who intervene in the affected zone must do so with no outside communication at all.

It is also believable that individuals in both the impacted and surrounding areas may give in to panic or even desperation in response to the direness of the condition. In such a situation one does not need much imagination to visualize the mobs of individuals who may, with doomsday, it's-the-end-of-the-world resolutions in mind, begin to thrust outward into nearby communities or towns, causing damage, looting, rioting, or looking for provisions to steal. Likewise it's easy to picture citizens being prepared to defend their property and livelihood by whatever means necessary. With authorities spread thin under such circumstances, one can easily see where this powder keg could quickly escalate out of control.

With masses exposed to such desperate conditions simultaneous to overextended authorities, the many possible responses on the part of the general populace are virtually unpredictable. However, history gives many examples that show us when crowds are left to their own devices under dire circumstances, it is more than plausible that they can overreact, panic, or otherwise become dangerous. The examples riddled through history are too numerous to cover in entirety within this book, but for argument's sake, we will list a few by type here.

The stampeding crowd

Many times when the concept of a stampeding crowd is mentioned, one has the vision of a person falling and others trampling on the person's body until he or she is dead. However, this is not usually how people are killed in a stampeding crowd. In fact it can be a much more subtle process. *Crowd crush* is a phenomenon where an individual within a crowd is squeezed by those around him or her until the person is asphyxiated, literally unable to expand the lungs enough to breathe. A more easily understood version of this is progressive crowd collapse, wherein one individual falls, knocking another next to him or her over, resulting in a domino effect that often kills those at the bottom of the pileup.[37]

Crowd crush is precisely what claimed the lives of ninety-six football fans in April of 1989 at the event that later became known as the Hillsborough disaster. At this event 10,100 fans filed into the small side of the Hillsborough Stadium, known as Leppings Lane, where seven small standing-room-only pens had been allocated to this crowd. Thirty minutes before the start of the game, just under half of the 10,100 fans had made it in the gate, with 5,700 still waiting to gain access to the small Leppings Lane pens, which stood at ground level and offered a good view of the field. Less than ten minutes before kickoff was scheduled, 2,000 fans squeezed hurriedly into pens 3 and 4, preoccupied with making it in time to see the kickoff and crushing those who stood closest to the gate facing the field. As the innermost crowd became excessively compressed, some escaped by climbing the tall fences to eject themselves from the pens, while others were lifted to upper spectator tiers by onlookers who could see the danger below. When part of the structure in pen 3 began to shift, it caused some people to fall, initiating progressive crowd collapse. In the end authorities determined that ineffective crowd control was the culprit in the event, with contributing factors from "police errors in planning, defects at the stadium and delays in the emergency response."[38]

In November of 2010, in Cambodia, 347 people were killed and more than 750 were injured at the Khmer Water Festival when the crowd on both sides of a bridge began to thrust inward, crowding those in the middle of the structure. Some jumped to the water below in order to escape, while others were compressed in the middle, causing their deaths. Hundreds of other terrified crowd members were killed by electrical wires pulled down by panicked individuals attempting to climb up the bridge's structure to safety.[39] In this case some blame the efforts made by police to control the crowd, stating that police fired a water cannon at the crowd, which caused the panic and initiated the stampede. Authorities, however, deny claims pertaining to the water cannon and of deaths by electrocution.[40]

The panicked crowd

Another bridge stampede took place in August of 2005 in Baghdad when a million Shiite pilgrims were making way to a shrine, which required crossing the Al-Aaimmah Bridge over the Tigris. The collective was already nervous, as there had been a mortar attack earlier that day, which had killed seven people. When one individual on the bridge shouted that a suicide bomber was in their midst, the crowd immediately went into panic. The gate to the bridge was open, but the structure itself was closed, creating a dead end and penning in the foremost crowd. Thus, as the throng pressed in, those in front had nowhere to go. More than 950 people were trampled, were smashed by the pressing crowd, or fell into the Tigris and drowned.[41] Ironically an assessment of this scenario led authorities to explain that there was no *actual* threat in this situation, stating that the rumor of a suicide bomber "caused chaos in the crowd, and the crowd reacted and caused this incident to take place."[42]

The angry crowd

In March of 1991 Rodney King was on parole for robbery when he evaded police via a chase across the streets of Los Angeles. Upon seizure of the man, police commanded him to climb out of his car, whereupon he was kicked and beaten brutally with police batons for a quarter of an hour while others (including other officers) watched without intervening on the man's behalf. Injuries sustained by the man included skull fractures, permanent brain damage, and broken teeth and bones. A little over one year later, when the verdict against four officers facing charges of "excessive use of force" came back "not guilty,"[43] the populace of the city was enraged and within three hours began to take justice into their own hands.

The first store to be looted was a liquor store, near where a trucker—the first victim to be pulled from his vehicle and beaten in correlation with this event—was left on the street bleeding. Soon rioting and looting (theft of both merchandise and cash registers, including money therein) were rampant, and random, hate-fueled thrashings were taking place all over the streets of south central Los Angeles. Gun and liquor stores that were looted of all their merchandise provided the already angry horde with means of impaired judgment and escalated violence, and more random attacks took place on the streets. Angry mobs turned their firearms toward the skies, shooting at newscast helicopters that hovered above. Cars were wrecked and/or set on fire and left in the street to burn. Traffic jams and barricaded streets made it impossible for residents to flee and in some cases for police or other authorities to get in. Some who attempted to evacuate the area made it as far as the airport and found that it had closed in response to the riots.

Meanwhile, due to the fires throughout the city, several power grids in the area were down. As a result many families in the area went without electricity. Initial, obvious speculations about this plight include the lack of electric heat or air conditioning, lights, and other electronic conveniences. However, most of the families in this area lived in small, bare-necessity-type apartments with no significant food storage features such as pantries or large kitchens, and they likewise went without refrigeration for days. Those who ran out of canned goods were forced to go to a Red Cross shelter or other facility in search of safe food to eat. When violence on the streets became so volatile that they did not feel they could leave their homes, some even went hungry, opting for safe starvation over risking traveling even mere blocks to a food bank or other community resource for assistance. Additionally some of the food shelters at that time, due to the rioting, resorted to having police escorts help them get supplies brought in, and those who were unable to obtain this assistance sometimes went without deliveries, making it impossible to meet the escalating needs they faced due to climbing numbers of hungry people seeking aid.[44] Few businesses remained open during this time, and those that attempted to maintain commerce while the riots continued were located smack in the middle of the violence, meaning that despite the fact that supplies may have been available, many people bypassed the pursuit of such goods due to fear. Armed forces standing guard were outnumbered and unable to curb the enraged populace, and the tumult continued.

The city imposed a dusk-to-dawn curfew, and reinforcements were called in to the tune of 9,800 National Guard troops, 1,100 marines, and 600 army soldiers during this five-day period. Arrests totaled almost 12,000 individuals, more than 50 people were killed, over 2,000 were injured, and the city sustained more than one billion dollars in damage and lost over one thousand buildings.[45]

While it is unlikely that a crowd experiencing a disastrous astral event would be as angry as that of south central Los Angeles during that time, it *is* possible that those who believe the end of the earth is near (and feel they have nothing to lose) might act in a similar reckless fashion. Even if random violence does not occur in the form of beatings, the potential is very real for similar conditions of looting and theft, which would be met with resistance, thus resulting in violence. In addition to these factors a cosmic event would have the added strain of the damage dealt by the natural disaster itself, which Los Angeles did not contend with at that time. A crowd that is angry, fearful, or desperate can, as NASA's statement on asteroid collision preparedness stated, possibly be the most lethal threat in this scenario.[46]

National Guard, FEMA, and Homeland Security responses

Hurricane Katrina

In August 2005 the Saffir/Simpson Hurricane Wind Scale category 5 Katrina swept across the southeast region of the United States, its 170-mile-per-hour winds dealing damage totaling approximately eighty-one billion dollars, killing more than eighteen hundred people, and displacing 1.2 million people from their homes. Retired army lieutenant general H. Steven Blum describes the response to this disaster by the National Guard as the "most massive military response to any natural disaster that has ever happened." Blum explains that the electrical grid was completely out and that during initial rescue attempts there was no communication available throughout the impacted area via telephone (landline or cell), internet, radio, or even television. Over the subsequent days fifty thousand National Guardsmen were dispatched to help with the rescue, which was concentrated in Florida, Alabama, Louisiana, and Mississippi—the regions hit the hardest. Roads leading into some areas were destroyed, described as "slabs and steps," causing delays for incoming troops, who took six painstaking hours to make the sixty-mile drive from their departure point to the heart of the damage in one of the affected zones. Facing a lack of communication and the desperate, immediate need of survivors, judgment calls were made by those moving in to assist, who had no time to await orders or backup. Due to failed levees surrounding the city, flooding in New Orleans escalated for four days after Katrina hit, leaving survivors on their rooftops, awaiting rescue, and even rendering the Louisiana Guard's center at Jackson Barracks unable to help,[47] along with other local official offices that would have otherwise provided relief.[48]

Katrina's fury left evidence along 6,400 coastal miles and covered a range of 90,000 square miles. Over 24,000 people were rescued, and 9,500 medical patients and personnel were evacuated by the coast guard during the subsequent flooding, while less-fortunate victims drowned. The wreckage was so pronounced that "the Coast Guard launched what would become the largest search and rescue mission in [the] nation's history," dispatching more than one hundred response vehicles, which assessed initial damage and then hunted for survivors.[49] National Guard troops faced with the sheer devastation of the impacted zone performed search-and-rescue duties, assisted with evacuation, and coordinated medical intervention efforts, along with dispersing provisions such as meal-replacement bars and safe drinking water.[50]

Despite the rally made by government agencies to assist in the wake of this terrible event, the Federal Emergency Management Agency (FEMA), a division of the Department of Homeland Security, has suffered criticism for its response during Katrina's aftermath. FEMA experienced a delay in responding to the request for help sent by Louisiana governor Kathleen Blanco to President Bush, petitioning federal assistance, due to bureaucratic hurdles in the process of such a request. Regional assistance was sent in a timely manner, but by the time assistance had been approved and was dispatched on a federal level, "the hurricane crippled many state and local emergency agencies in Louisiana, Mississippi, and Alabama leaving them unable to respond without federal help." In other words, the local offices were impacted by the storm and unable to conduct optimal efforts on their own behalf. Outside, federal resources were needed earlier than they were dispatched. As a result rescue helicopters, safe drinking water, food, and medical supplies that had been requested in advance arrived late, a fault that

Connecticut senator Joseph Lieberman stated "allowed much more human suffering and property destruction to occur than should have."[51]

Considering that Governor Blanco's first request for federal intervention was sent on August 26, 2005, and Katrina's full wrath struck the region early in the day on August 29,[52] one would have hoped that enough time was allowed for emergency preparations to be made. However, that is not what happened. Granted, FEMA has undergone many changes since (and because of) these circumstances, but it still remains that the warning time for an asteroid event could likewise be as short as several hours, if the asteroid is small enough and comes from the daytime sky—the precise reason that the Chelyabinsk event caught the scientific community by surprise.[53] And calling the asteroid small is subjective; recall that this particular asteroid was sixty-five feet in diameter,[54] a measurement considered minuscule by some experts, and yet when it hit the earth at 40,000 miles per hour near a Russian town, it was large enough to release five hundred kilotons of energy,[55] injure more than fifteen hundred people in the community, and damage thousands of buildings.[56] Bearing this in mind—added to the understanding that an astral event could unleash many unpredictable components such as tsunamis, hurricane winds (with subsequent flooding), cloudy plumes which block the sun's light, or even nuclear/chemical debris—one has to wonder if there could possibly be enough personnel in place to adequately prepare for all the needs of a populace under such circumstances. Furthermore, some examples of crowd mishaps listed in this chapter show that in some cases intervention by authorities was misread by an already tense crowd, causing matters to escalate. Thus, in addition to the emergency and all it entails, the crowd begins to panic or grow desperate, and responders could find themselves facing the most challenging response, search and rescue, and crowd control mission the world has ever known. Just imagine what is going to unfold when Wormwood strikes Earth, and, among other things, one-third of drinkable water is gone in a day.

Chapter 5

HIS NAME IS WORMWOOD

A<small>ND THE THIRD</small> angel sounded, and there fell a great star from heaven, burning as it were a lamp, and it fell upon the third part of the rivers, and upon the fountains of waters; and the name of the star is called Wormwood: and the third part of the waters became wormwood; and many men died of the waters, because they were made bitter" (Rev. 8:10–11).

In the final book of the Bible the apostle John is well into a lengthy, colorful description of coming events. Elderly and exiled, John in this vision is an eyewitness as a series of seven seals on a vellum scroll are peeled back, one by one. Only the Lamb of God who was slain was deemed worthy to open the seals of what is obviously a very important document.

Seals one through six have already been opened with dramatic and effective results. Revelation 8 starts by telling of a period of time—one half hour—that heaven will be completely silent after the breaking of the seventh seal. Think about that...In a place where worship is nonstop around the throne, suddenly there is total silence for *thirty full minutes,* like pounding heartbeats, slowly ticking away in anticipation of what is about to happen. Then seven angels, appointed for this specific moment in time, prepare themselves to sound (v. 6).

THE SIGNIFICANCE OF SEVEN

One does not get far into Revelation without noticing a few interesting numeric patterns. Apart from the frequent *one-third* seen throughout the trumpet judgments, the number seven is probably the most obvious, and Revelation is not unique in this regard. Just as a quick aside, it is difficult to appreciate any depth of Messianic study without an understanding of the fullness and profundity of the number seven in the Bible. In the first chapter of Revelation alone the reader sees that letters are written to seven churches, seven spirits are described before the throne of God, seven golden candlesticks are mentioned, and there are seven stars in the hand of the Lord.

In a biblical context seven is a number connected to divine perfection (Gen. 2:2; Ps. 12:6), spiritual completion (Rev. 1:4; 3:1), and blessing (Gen. 2:2–3), just to name a few. Indirectly we can find examples of multiple links to the number seven in one symbolic circle: most people are familiar with the creation story and know that God created Earth in six days, declared it good, and then rested on the seventh day. The number seven is incredibly significant in Bible study (many refer to it as "the Lord's number"), and its first mention as early as His day of rest gives some amazing hints about the symbolism and importance. On day six of creation God brings forth each living creature after its own kind, thereafter designing humankind and establishing their dominion over the earth, followed by the first mention of Sabbath rest, occurring on the seventh day. Many people may make the tongue-in-cheek comment that God rested because He was tired, but this is not the case. God is all powerful. The mention of rest does not indicate that He slept; in fact Psalm 121:3–4 states that He never slumbers. Thus, the "rest" referenced in this event was in connection with the fact that God saw that His creation was "good" (Gen. 1:4, 10, 12, 18, 21, 25, 31). Everything He had desired for His creation to be sat before Him. Everything was *divinely perfect, complete,* and totally *blessed,* and God paused to enjoy His handiwork.

VARYING POSSIBLE INTERPRETATIONS

Whereas thus far our attention has been on asteroids, we now turn our attention to other ways of interpreting this Wormwood picture. Regardless of how one chooses to take this description of trumpet activities, every reader should agree there are horrendous events on tap within the vision, accompanied by extremely serious consequences. These end-time passages are laying out a progression of global cataclysms of undeniable biblical proportions.

Pete McCutcheon—our protagonist from the fast-paced, fictional opening to this book—held a theory that is shared by other experts far less fictional. And while not everyone asserts that prophecy will be fulfilled *specifically* via an incoming comet or asteroid, admittedly a casual reading of the entire chapter of Revelation 8 certainly lends itself to such an interpretation.

That said, there are other options regarding the meaning of this Wormwood and what its arrival may portend in this prophetic trumpet discussion. Often, prophetic passages such as those predicting the famine, pestilence, and earthquakes in the last days, referenced by Jesus in Matthew 24, offer clarity, while others remain somewhat fuzzier, as "through a glass, darkly" (1 Cor. 13:12), to be revealed at a later time of appointment, thus left open to speculation and varying interpretation.

Many Christians—and even churches—avoid studying the Book of Revelation, often believing it to be too confusing, inapplicable to their own lifetime, exceedingly terrifying, or bafflingly metaphorical. However, individuals who avoid this book unknowingly shortchange themselves; readers are told in the first few verses that those who read and *truly hear* this revelation will be blessed. Furthermore, believers are challenged in 1 Thessalonians 5 not to be found ignorant regarding these end-time events because we do not belong to the night or in darkness. On an additional, practical note, for those who are alive on this earth when such prophetic events *do* begin to occur, happenings may seem less ominous and overwhelming since they have already been disclosed by the One who has "overcome the world" and everything therein (John 16:33).

TRUMPETS

Although this book is absorbed with the third trumpet, it stands to reason that all trumpet blasts (occurring in Revelation 8–9) should be visited briefly as a reminder of what else is expected to occur around that time and in what order.

Remember that if the apocalyptic results of these events are literal (destroying enormous portions of the earth and its inhabitants, filling waters with blood, darkness, and so on), then it stands to reason to interpret the judgments as literal also.

> **First trumpet (Rev. 8:7):** All green grass is burned, as well as one-third of the planet's trees, when "hail and fire mingled with blood" were "cast upon the earth."

Second trumpet (Rev. 8:8–9): Something enormous, only described "as [if] it were a great mountain burning with fire," is thrown into the sea (salt) waters. This destroys one-third of all ships, kills one-third of all ocean life, and turns one-third of the water into blood (just like the plague of the Exodus narrative). The "great mountain" in this trumpet judgment is often assumed to be a flaming asteroid as well.

Third trumpet (Rev. 8:10–11): A "star" called Wormwood falls to the earth, making "bitter" one-third of all the springs and rivers and therefore poisoning many who drink from them. This trumpet judgment is similar to the last, but instead of altering the salt water, it affects the non-oceanic (drinkable, potable) fresh water.

Fourth trumpet (Rev. 8:12): One-third of the sun, moon, and stars are struck and destroyed, resulting in one-third of each day submerged in total darkness.

Fifth trumpet, or first woe (Rev. 9:1–11): Here we encounter a clear personification. Yet another "star" falls from the sky, but this time it is clearly a *he* and not an *it*. He is given the keys to the bottomless pit (or abyss), and once it is unlocked, the smoke that rises up darkens the sun and air. Powerful locusts (scholars believe these to be demons/fallen angels, as they answer to Abaddon/ Apollyon, the "destroyer") emerge from the smoke, instructed only to "torment" (to "strike") but not to kill the people who do not have "the seal of God in their fore- heads" (those mentioned in Revelation 7:1–8). The locusts will be completely harmless to all plants and vegetation, but to the people they have been sent to torment for "five months," their description is truly terrifying (shaped like horses, wearing crowns and armor, faces like men, hair like women, teeth like lions, tails like scorpions, and the buzz of their unholy wings so loud they will be "as the sound of chariots of many horses running to battle"). These entities will bring such misery that "in those days shall men seek death, and shall not find it." (Somehow even suicide will not be an option; I have my own suspicions that our transhumanistic sciences will have something to do with achieving this kind of mandatory immortality, as I have written in previous works.) Of all the trumpet judgments, this one is so disturbing that it brings a deeper understanding of Christ's personal warning: "For then shall be great tribulation, such as was not since the beginning of the world to this time, no, nor ever shall be" (Matt. 24:21).

Sixth trumpet, or second woe (Rev. 9:13–21): Four (fallen) angels, bound by the Euphrates River, are released to wreak havoc with an army of two hundred million heavily armored warriors on horses with "the heads of lions; and out of their mouths issued fire and smoke and brimstone... [and whose] tails were like unto serpents," (Rev. 19:17, 19) killing one-third of the world's remaining population.[1] (Mathematically, between the fourth seal [Rev. 6:7–8] and this sixth trumpet judgment alone, half of the people on the earth have been wiped out.) Though great death and much fear result because of these bizarre warriors' attack, people still give themselves over to idolatry and moral depravity.

Seventh trumpet, or third woe (Rev. 11:15–19): Voices in heaven shout, "The kingdoms of this world are become the kingdoms of our Lord, and of his Christ; and he shall reign for ever and ever." The twenty-four elders of the throne room fall on their faces before God in worship. The temple in heaven opens, and the ark is visible. Lots of noise follows with voices, thunder, earthquakes, and great hail.

WARFARE IN THE SKY?

The trumpet judgments...What in the *otherworld* was going on in space at that time? The astral catastrophe discussed in chapter 4 reminds us that if Wormwood *is* an asteroid, its flight path to destruction is *possibly* related to some kind of space war between holy and unholy entities and deities. If that past event concluded with such massive levels of irreparable damage to our universe, and if it *was* related to a battle in the unseen realm, then what kind of ravaging will the earth be facing if the trumpets and seals of Revelation are similar?

Wormwood, then, would not be a *global* issue; it would stem from a *universal* issue. Could this potentially involve entities engaged in spiritual warfare within the universe, causing another round of planetary ping-pong? It becomes obvious through the descriptions of the trumpets (particularly numbers five and six) that a battle between good and evil is heating up.

Often in our corporate church imagination the prevailing narrative regarding the blowing of the trumpet usually spawns visions of a spiritual entity (angel), melancholy, but resolved, blonde with a white robe and feathery wings, perched peacefully on a cloud, awaiting his cue. When the time has arrived for him to blow the third trumpet and usher in the next judgment of God upon the wicked, he somberly lifts the modern brass instrument to his unwavering lips. After his blast quiets, he watches, disengaged, from his cushy seat in the heavens while a giant rock splashes into the earth's most important freshwater source.

But maybe, *just maybe*, there is an angel who is more involved than this cartoon version…and perhaps we're not looking at an impact/collision event at all.

With so much talk up to this point regarding asteroids, comets, stars, planets, satellites, planetary defense, and future space economies that inspire flashbacks of episodes from *Babylon 5*, it's easy to keep our collective sights only on "natural space activity" (however erratic) and forget *another* theory about Wormwood, which may, according to some scholars such as Michael S. Heiser, be the most obvious conclusion regarding these poisonous waters.

Could Wormwood be a *fallen angel*?

The notion that trumpet three, Wormwood, would be an *entity*—especially with how many spiritual/demonic beings these judgments are increasingly producing around this time—will probably appear in that day to both Christ's devoted ones and the Antichrist's followers as a real possibility, perhaps even the obvious conclusion. And if language that sounds as if it's describing an asteroid or comet is really referring to an evil spirit, then what *other* trumpet-judgment language might we be getting wrong?

Might all these details as revealed to John at Patmos be pointing to an *angel of famine*?

Well, that would certainly change everything. And I assure you, we will get to that. However, early on in that thread one bizarre planet that's not really a planet needs to be brought up and knocked down so we can proceed into the fallen-angel theory with a better sense of understanding.

NIBIRU

One popular fringe theory dominating this Wormwood affair is the idea that ancient Sumerian cuneiform tablets tell of a day when a mysterious planet (Nibiru) inhabited with gods (which folks today interpret to be aliens or ancient astronauts) will crash into the earth.

These gods are the "Anunnaki"—powerful, human-fate-controlling deities of the Sumerian/Babylonian pantheon—who landed on Earth, experimented on apes, and created us humans via genetic manipulation so we could mine precious metals for them. Though the purpose of the Anunnaki in this Mesopotamian angle is different from Genesis 6 or Enochic literature (which addresses the DNA manipulation the fallen angels carried out by sleeping with human women for the purpose of defiling the messianic bloodline), these two stories are synonymous enough that they are frequently seen as different cultures' interpretation of the same events. Therefore, if the Anunnaki/fallen angels are currently chilling out on a planet in a trajectory for collision with Earth—and if *fallen angel* terminology proves to be relevant to the Wormwood star of Revelation—then it stands to reason that so many would believe Wormwood to be Nibiru. It's appropriate, then, to begin this study by establishing whether the Anunnaki home planet Nibiru is, or is not, significant to any potential association with Wormwood and/ or fallen angels.

First and foremost, a clarification of terms…

Why this book will avoid reference to Planet X

Nibiru is, according to the Enuma Elish (the mythical Babylonian creation account, circa 1750 BC), the name of the star of Marduk, who is the creator god of the Babylonian pantheon, and chief god over all others. Translated from the pre-Aramaic, Eastern-Semitic language of Akkadian, Nibiru means crossing over, like the transition point of a river or stream. Within the context of astronomical space bodies, especially as it relates to circulating Sitchin-theory discussions about Wormwood and/or the Book of Revelation (discussed later), it has been associated with a crossing over and into Earth's atmosphere (as in a potential impact event).

Often connected to Nibiru theories is so-called Planet X, which can mean a number of things. To begin, the X means unknown, undiscovered, or unnamed, as opposed to the Roman numeral for *ten*. From Percival Lowell's use of the term in the early 1900s during his hunt for a ninth planet beyond Neptune (Pluto hadn't been discovered yet) that would account for the bizarre orbit of Uranus, to today's scientists' theories regarding a planet beyond Pluto, Planet X has simply been a stand-in name applied to mysterious space bodies eluding the reach of our current space imaging technology. So whereas Nibiru could be considered *a* Planet X—one among oh-so-many potential X bodies as long as we don't know what dwells beyond our solar system—that doesn't mean that when the interchangeable term Planet X is dropped into conversation, it automatically refers to Nibiru.

Therefore, in the interest of specificity and to avoid confusion with the numerous books and articles and news coverage referring to completely unrelated space topics, this book will proceed without reference to the ambiguous Planet X.

Influence of Zecharia Sitchin

Zecharia Sitchin (1920–2010)—of all things an *economics* major—was a self-proclaimed master of the ancient Hebrew, Akkadian, and Sumerian languages who taught himself how to read the cuneiform tablets left behind by Sumerians. According to his most popular and widely talked about doomsday theory, alluded to initially in his 1976 book *The 12th Planet*, Sitchin claims that the tablets spoke of another planet beyond our solar system called Nibiru. This planet (or "star") allegedly "crosses over" into our solar system every thirty-six hundred years. Spoiler alert: if his theory were to be true, the connection could easily be made between Marduk's pet planet and the apocalypse of Revelation. Of particular interest to *us* would be its association with Wormwood.

Though Sitchin personally steered clear from attempting to nail down a specific date of the collision/impact of Nibiru into Earth's atmosphere (except once in his book *The End of Days*, which gave a rough guesstimate of the year 2900), countless theorists (other self-proclaimed experts) have stepped up to do so—including ZetaTalk's Nancy Lieder, who claimed Nibiru would destroy human life in May of 2003, and David Meade (who calls himself a "Christian numerologist"), who warned that September 23, 2017, would be the end of the world.

These dates came and went without anything particularly interesting happening in Earth's atmosphere. That's the *short* version.

Perhaps the main reason (of several) that we won't get into every detail of the longer version of Sitchin's theory (involving all sorts of connections to aliens/early astronauts) is that it's a structure built on sand.[2] Sitchin lacked any formal education related to ancient languages and cuneiform writing...and teaching oneself how to read the writing of one of the world's most mysterious ancient languages without a shred of educational background in any related area of study is not necessarily the most reliable platform upon which translations should be produced. This would thereby result in his translation of a Sumerian Armageddon scenario being a highly unstable one. So far, though theories and postulations by scientists and astronomers attest to the very real possibility that there is another planet orbiting beyond Pluto, nobody is saying that *anything* crosses into our solar system every thirty-six hundred years, specifically, or that there is any legitimate correlation between that potential planet and the mythical Babylonian creation epic.

But in case this isn't reason enough to avoid a deep dig into Sitchin's works, consider this also: Sitchin ignored what well-known and well-documented scholarly research already existed regarding the astronomy symbolism/iconography present on cuneiform tablets and seals. Where Sitchin saw the Sumerians depicting eleven planets and alluding to a twelfth, reliable scholars of Mesopotamian history with a plethora of formal education and experience in archaeological discoveries see the Sumerians' innocuous drawing of the stars in similar style to other images they left behind. One such example is the VA243 cylinder seal, now housed in the Berlin Near Eastern Museum, which Sitchin famously claimed as showing "the sun in the center (!), surrounded by eleven orbiting bodies that include the Moon, Pluto and the yet to be recognized Planet X—Nibiru."[3] In order to come to this giant leap of a conclusion, Sitchin had to blaze straight past all established, scholar-reviewed evidence that clearly showed this bundle of sun and planets to be mere stars, done in the same likeness as numerous other Sumerian examples from the same people and in the same era. That is apart from the fact that the Sumerians never once, on any other tablet or seal or work of any kind, alluded to the fact that they knew about more than five planets, so to suddenly show on only this one seal that they knew of eleven appears to be inexplicable conjecture.

Sitchin's botched translations didn't limit themselves to only incorrect interpretation of Sumerian artwork. Words and sentences from the tablets are subject to completely inaccurate leaps as well. The works of many scholars—one such example being the renowned lexical genius of Sumero-Akkadian vocabulary Benno Landsberger—are disregarded each time Sitchin "translates" what these ancient tablets say. This leads to such errors as the claim that the tablets identify the planet Nibiru as the home of the Anunnaki. Not surprisingly the cuneiform tablets, according to qualified scholars and historians, say no such thing. In fact they don't recognize Nibiru as inhabitable at all.

The list goes on for why I don't feel it's necessary to go much further on Sitchin's theories about the end of the world. Moving right along…

Then what *is* Nibiru?

At the opposite end of the educational spectrum from Sitchin, my friend and scholar Dr. Michael S. Heiser has an extensive graduate résumé and *decades* of experience in careful, accurate, peer-reviewed translation of ancient languages. His academic degrees require a full paragraph to unravel—University of Wisconsin-Madison, PhD, Hebrew Bible and Ancient Semitic Languages; University of Wisconsin-Madison, MA, Hebrew and Semitic Studies; University of Pennsylvania, MA, Ancient History (Israel, Egypt)[4]—and his list of academic honors is even longer. On his record of relevant graduate courses sits language studies of all kinds, including Akkadian and Sumerian, as well as an impressive stack of education in archaeology and ancient history. (He also happens to be the most well-known challenger of Sitchin's work, and I remember well when, before Sitchin passed away in 2010, Heiser challenged him to a debate, which went unanswered. Heiser's website, known by the provocative domain SitchinIsWrong .com, works methodically to expose the errors of the self-taught cuneiform translator.)

At the end of the day, because the cuneiform tablets clearly identify Nibiru as "a planet (specifically, Jupiter, but once as Mercury), a god (specifically, Marduk), and a star (distinguished from Jupiter)," leading to what Dr. Michael Heiser calls the "tri-fold (fourfold if you count Mercury) designation for Nibiru,"[5] landing at an absolute, concrete conclusion about what Nibiru is, exactly, is difficult, if not impossible, for even the best of scholars. (Though Heiser goes on to note that the tablets most certainly do *not* tell of a planet past Pluto on a thirty-six-hundred-year return to our solar system, because the Sumerians recorded Nibiru sightings every year, as an *annual* event.) However, the following educated guess about what Nibiru exactly is agrees with all evidence and makes complete sense in the context of the ancient documentation. What we *can* piece together about Nibiru from the cuneiform tablets is the following, in part:

1. It serves as a "shepherd" star, guiding and supervising the course of all other stars (Enuma Elish Tablet V, line 6; Enuma Elish Tablet VII, lines 130–131).

2. It marks the "crossing place" between heaven and earth (Enuma Elish Tablet VII, line 124; also remember that "crossing place" is what the word Nibiru means).

3. It "divides the sky in half" (Astrolabe B, KAV 218B ii, lines 29–32; Mul.Apin I.i: 36–38).

4. It has been described as a "red star" (Astrolabe B, KAV 218B ii, lines 29–32), whose "light is dim" (Mul. Apin I.i: 36–38).

5. And perhaps most odd: "It keeps changing its position and crosses the sky" (Mul.Apin I.i: 36–38).[6]

On their astronomical tablets the Sumerians alluded to their knowledge of five total planets (never eleven) and documented the celestial objects in the sky based on the fixed positions of the stars. These stars were then observed daily, weekly, and monthly in relation to where they sat on the horizon, farther north vs. south, and so on, and these locational patterns were named after the gods Enlil, Anu, and Ea. Further, the path of the planets, sun, and moon were carefully documented in addition to the phases of the moon, varying durations of daylight hours, and changes in weather patterns.

It's quite impressive to see how much they recorded.

Then, anomalously, Marduk's Nibiru—confusingly referred to so far as four different things (star, god, Jupiter, Mercury)—at some point (and perhaps on more than one occasion) changed its location in the sky, at least as far as the Sumerians *perceived* when they tracked the object from Earth.

What does all this add up to?

In the Enuma Elish Mesopotamian creation epic, Tablet VII, line 124, Nibiru is given yet another identifier: Marduk *himself*.[7]

Throughout the tablets the name Nibiru is given what Heiser

is calling a "DINGIR"[8] sign, which associates a word with a deity, meaning that Nibiru is as much a god as a star, planet, or anything else. So when Marduk is here *directly* referred to as Nibiru, this language—regarding the shepherding sky divider that apparently moves around all over the place—starts to make more sense. All signs point to the likely conclusion that descriptions of Nibiru are all personifications of the Babylonian almighty creator-deity, Marduk. And if that is correct, then most of these erratic star behaviors—as perceived and documented by the ancients—are being carried out by a god who, by his own prerogative, could direct the stars however he wants, divide the sky in whatever way he sees fit, and blink whatever colors in whatever brightness and from whatever location in the sky he feels like gracing his presence with.

Once and for all, it is bad exegesis of these ancient cuneiform tablets to suggest that Nibiru is a planet inhabited by ancient astronauts that are also fallen angels and that this planet will collide with ours. And since this is a reasonable conclusion, then it dismisses the idea that Wormwood is Nibiru.

WORMWOOD'S FALLEN ANGEL/FAMINE CONTEXT

With the concepts of Nibiru as a planet out of the way, and now that we no longer need to fear the wrath of the Anunnaki's fateful collision, let's take a look at what Revelation might actually be saying about a fallen angel/star in proper context.

But first, a quick trip down Biblical Interpretation Lane...

Revelation: literal or not?

When discussing whether the falling star could actually mean a falling angel, the conclusion based on so many supporting scholarly materials is an easier one to come to than the other, which suggests that the bitterness (poisoning) of wormwood actually means famine and has nothing to do with water. Nevertheless *both* concepts demand that the question of literality in the interpretation of Revelation is at least visited briefly so readers can understand 1) why the interpretations—even among scholars—vary so widely, and 2) why many of these theories are equally as plausible as several of the others. With that in mind, I'll keep it short.

There is a popular (and respectable) principle of biblical interpretation that many instructional books (including college textbooks in the most reputable Christian universities) teach as one of the first rules to follow when studying the Bible. Though the wording may change a bit on a case-by-case basis, the gist goes about like this: "If a verse *can* be interpreted literally, it should be. The only occasion when a scripture should *not* be taken literally is when doing so creates an absurdity."

For instance: "The cloke that I left at Troas with Carpus, when thou comest, bring with thee, and the books, but especially the parchments" (2 Tim. 4:13). This is a clear, nonallegorical order for Timothy to bring Paul's cloak and documents. It comes at the end of a didactic letter with instructions for ministry, not within the frame of poetry or apocalyptic, and so on. In other words, to take 2 Timothy 4:13 as literal *does not create an absurdity*, because it's not illogical or unreasonable in any way to assume Paul wanted Timothy to swing by with some supplies next time he was headed out Paul's way.

However, "except a man be born again, he cannot see the kingdom of God" (John 3:3). This statement, taken to the fullest application of literality, means that a man must crawl through his mother's womb again before he can go to heaven. But as we know, even Nicodemus, the recipient of these words as Christ spoke them, realized that a literal interpretation would *create an absurdity*, so he sought clarification, and Jesus further explained that this rebirth was "of the Spirit" (John 3:4–6), leading us today to understand that this was a metaphor, a symbol of a spiritual change into what is elsewhere described as a "new creature" (2 Cor. 5:17).

Therefore, to follow one of the most fundamental principles of biblical interpretation, wormwood would be a poisoning of fresh waters (not famine), and Wormwood would be a space-body mass object of some kind (not an angel or fallen angel). Assuming we're dealing with asteroids and a poisoning much like the Lake Nyos tragedy (see the next chapter), we have been able to conclude a literal interpretation that, once again, does not create an absurdity. This is perhaps why many scholars of the Word resist the allegorical or metaphorical approach to the Book of Revelation any time we *can* see a logical, literal explanation of what's coming.

Yet the very first rule of biblical interpretation, which *must* be acknowledged before any and all others in any serious approach to Scripture (and any dependable literature in this area will agree), is this: There can only be *one true* meaning of the verse in question, and that is the meaning that the author of that book intended for his original readers. No interpretation that disagrees with the author of the book (and by extension the Holy Spirit, who led such a composition) can ever be the right one. The real issue lies in getting to the bottom of what *that* author meant and not in deciding what we think makes the most sense to us today. That faulty approach is called *eisegesis*—the putting into Scripture what was never there—which is the opposite of *exegesis*, which means the pulling out of Scripture what has been there from the beginning. It's also to commit the error of placing hermeneutics (the what-it-means-to-us-today task) ahead of exegesis (the what-it-meant-to-them-back-then task).

And yet this conundrum has never been bigger than when we get to Revelation, for the following reasons:

1. Unlike a didactic work or a historical narrative, John at Patmos was documenting a *future* reality,

not a past one, and the language of his vision often demands allowance for the same kind of prophetic imagery we have seen elsewhere in the Word, ripe with symbolism and metaphor, much like some of the messages given by the major prophets of the Old Testament.

2. Parts of Revelation can be interpreted literally without creating an absurdity (such as all the asteroid talk we've covered thus far), whereas other areas absolutely cannot—such as the idea that the "great whore" called "Babylon" was an actual "woman" with "seven heads and ten horns" and held a cup filled with "abominations and filthiness of her fornication" as she is "drunk" on the "blood of the saints... [and] martyrs of Jesus" (Rev. 17:1–6), and this list only goes on.

3. As will be shown shortly, the original audience can and did recognize a language and imagery (star as angel) from first-century authors that are lost from mainstream Bible studies today. This unfortunately tends to leave the interpretation of a difficult book such as Revelation only to those supremely educated scholars whose field of specialty (like Heiser and Gregory K. Beale) lies in comprehending the full and complete culture at the time of the original author's penning. Since this is a minority, that means that an extreme minority of the church is sorting it all out, and the responsibility of comprehending literal versus nonliteral appears insurmountable for many.

Nevertheless, if we *do* allow such think tanks of the church to have their say, we find ourselves challenged to follow the trail that suggests *all* these trumpet judgments are symbols and *none* of them are literal, because *the only true meaning of the verse in question is the one the first author intended.* If John of Patmos intended to share his already overwhelming prophetic vision in language his contemporaries would understand, perhaps because what he saw was already so hard to put into words of *any* language, then we too should be willing to visit that possibility. It all boils down to context.

Star as angel?

"What exactly is the proper context?" you may ask. It's what Heiser refers to as a "shocking proposal" in one of his YouTube videos on the subject:

> The original audience for the book of Revelation...would have had a very different understanding than this idea that the earth is gonna get hit with some sort of planet in the future. As we begin to think about this a little bit more, I have a shocking proposal. Rather than relying on claims of ancient aliens to interpret the Book of Revelation, it might be wiser to read the first-century book in light of its first-century historical and literary context. I know that's radical, but that's what I'm gonna propose.
>
> As it turns out, there is, indeed, another way to read that passage in the Book of Revelation and to understand the meaning of Wormwood. In ancient literature, in the centuries preceding, and contemporary with, the New Testament, falling stars were a familiar way, I would say a *stock* way, to describe the rebellion or descent into sin of heavenly beings. You might wanna consider them fallen angels. This is because people in antiquity thought of celestial objects in the sky as *angels*, or *gods*, or animated in some way. The stars changed positions each night; the moon did as well. The sun moved across the sky, and it created the impression to the ancients that they were living beings.[9]

Well known among Bible scholars, but largely undiscussed in lay Christianity, is the fact that the word *star* (Greek *aster*, occurring twice in Revelation 8:10–11) is frequently a personification of either a saint or an angel in the Word, as well as classic and ancient extrabiblical Jewish writings or apocryphal accounts (cf. 1 Enoch 43:1–44:1; 104:26; 2 Baruch 51:5, 10; and several others). Certainly there are countless examples of similar treatment of terminology outside the Word, such as the Marduk star actually representing a god. One major *scriptural* point of comparison scholars also make is the link between Michael as guardian angel in the Old Testament Book of Daniel 12:1 and his relation to the intelligent and personified stars (Hebrew *kowkab*) two verses later.[10] This association was so well known by the New Testament writers that it felt like a natural extension of terms by the time John of Patmos wrote that "the seven stars *are* the *angels* of the seven churches" in Revelation 1:20 (emphasis added). The English word *are* here is from the Greek *eisi*, literally only meaning "are, be, were, etc.,"[11] a direct identifier, and there can be no confusion. From the *Theological Dictionary of the New Testament* we see repeat personifications of the Greek star as angelic beings:

> *astēr* denotes a star....The ancients regarded stars as "beings" and even deities, but in the OT [Old Testament] they execute God's command and declare his glory (Is. 40:26; Ps. 19:1)....Apocalyptic speaks of stars falling from heaven (Mk. 13:25; Rev. 6:13) or being obscured (Rev. 8:12). A falling star may have destructive effects (Rev. 8:10–11)....The morning star of Rev. 2:28 has been understood to be the Holy Spirit, the chief stellar angel, or the dawn of salvation; in 22:16 it seems to be Christ himself. The star which appeared to the Wise Men accords with messianic expectation on the basis of Num. 24:17 [declaring the star to be a "he"], but what star it was, and how the wise men interpreted it, we cannot say for certain.[12]

And elsewhere, in *The New Bible Commentary*, the close proximity of the *other* stars of the trumpet judgments as angelic messengers of death appears obvious enough to immediately consider Wormwood an angel as well: "Since the star that falls at the sounding of the fifth trumpet (Rev. 9:1) is an angelic being, it is possible that Wormwood is also an angel."[13] The association is strong enough that it can be found listed in scholarly works all over the place, with *The Book of Revelation: The New International Greek Testament Commentary* by Gregory K. Beale likely being the most trusted (and exhaustive) source regarding the Book of Revelation.

Beale, a professor of New Testament and biblical interpretation at Westminster Theological Seminary and the author of over twenty critically acclaimed theological works, is considered by many fellow scholars (including Heiser) to be a leading expert in the breakdown of Revelation. His writings appear in the bibliographies of *hundreds* of prominent reference books, including Bible dictionaries and encyclopedias whose editors considered Beale's opinion both unbiased and factual enough to be featured as their mainstream explanation behind countless biblical topics.

Beale's take on the "angel star" named Wormwood notes that "the observation that the descent of the burning mountain in [Revelation 8:10] is parallel to the descent of the burning star in v 8 also indicates that the star should be identified as an angelic representative of an evil kingdom undergoing judgment. Here the judgment of Babylon's angel is in view, since v 8 concerns the judgment of Babylon the Great. The identification of the star as Babylon's representative angel becomes more convincing if v 10 is understood as alluding to Isa. 14:12-15."[14]

Of course the Scripture reference of Isaiah that Beale mentions here is the most cited verse in all the Bible whenever Lucifer's fall account is brought up or questioned: "How art thou fallen from heaven, O Lucifer, son of the morning! how art thou cut down to the ground, which didst weaken the nations! For thou hast said in thine heart, I will ascend into heaven, I will exalt my throne above the stars of God: I will sit also upon the mount of the congregation, in the sides of the north: I will ascend above the heights of the clouds; I will be like the most High. Yet thou shalt be brought down to hell, to the sides of the pit." So just to be clear, according to Beale's interpretation, the star of Wormwood is a personification, an angelic agent of evil Babylon—which is, itself, a symbol of the Antichrist government in power during the Tribulation—not an asteroid, comet, or space body of any kind in trajectory toward Earth (and *especially* not Nibiru). The association Beale makes to the angel as simply a "representative" leaves it open-ended as to whether the angel is a satanic messenger or a heavenly one, though Heiser and others believe understandably that this entity is evil, and essentially the results would be the same either way, since God is allowing the judgment to take place.

And speaking of Heiser, let us steer back to his opinion and the "shocking proposal" context of how this literature would have come across to first-century readers for a final layer of confirmation:

> Original readers of Revelation's Wormwood description would therefore quite readily have been thinking of fallen angels. So, to read a description of Wormwood as a falling star come to Earth, to *them*, would have suggested a supernatural rebellion, or some kind of unleashing of supernatural, hostile powers of darkness. They would not have been thinking of comets or meteors or a rogue planet, for sure. The point is that Wormwood may be a spiritual event or concept, not a literal, astronomical one....
>
> Now, [wormwood] is the name of a bitter herb. It's interesting to note that in *all* of the texts and records of ancient Greco-Roman astronomy—I'm gonna say that again—*all* of the texts and records of ancient Greco-Roman astronomy that have survived to this day, and there are a *lot* of them, in *all* of that material, there isn't a single instance where any actual star

in the heavens was called absinthe [or wormwood, *artemesia*, etc.]. None of them. Now this suggests very strongly that the ancient writer was *not* thinking about an astronomical object when he used that particular term that we translate as "wormwood."[15]

(Worthy of note: Heiser is not here making the statement that wormwood or Wormwood have anything to do with a literal herb. It couldn't be *that and* an angel, and we already showed in our study on etymology in the previous chapter the fallacy of that assumption. His emphasis is on showing that the Greco-Roman astronomical records never assigned *wormwood* or any of the associated herb terms to any other space star in history.)

To add another layer of evidence to this approach, the first-century readers would have also been familiar with the widely circulating and heavily referenced Book of Enoch, and it is even mentioned in our canonical Bible (Jude 12–15). In lieu of that, the Book of Enoch 18:13 and 21:3 make a strong and *startling* connection, not only in the imagery of angels as stars but also to their appearance as great burning mountains as they fell from heaven. Consider how many instances we run across "stars" and "fire" and "mountains," and so on, during the trumpet judgments!

But another question is now presented: If the star is really an angel and not an asteroid that might set off eruptive phenomena or some equivalent, then what *exactly* are the "bitter waters" that result in death?

As shown already, when we trace *wormwood* back to its Hebrew root, *la'anah*, it represents extreme suffering, destruction, the opposite of righteousness and justice. That could mean just about anything in an eschatological trumpet judgment creating mass death. The connection mentioned at the beginning of this chapter in relation to famine is peppered here and there throughout scholarly studies, but once again, and not surprisingly, we find ourselves redirected back to Beale for a more precise insight.

Bitterness as famine?

Beale postulates that Wormwood may not be referring to a literal poisoning of fresh waters but a pouring out of bitter suffering, a *curse*, that results in death by *famine*. The acceptance of this theory requires letting go almost entirely of any literal space-body interpretation, thrusting us into a complete makeover of all we've said about asteroids or the astronomical universe up to this point and requiring our minds to open up a bit more than some would like—though stretching is not always a bad thing, so let's tear into it.

To begin, *fire* isn't even remotely limited to literality, and even some occasions when it has been a literal flame, it has also been a personification—though technically, in cases such as the burning bush (Exod. 3:2, 6), the pillar of fire (Exod. 13:21), and the Lord's descending upon Mount Sinai (Exod. 19:18), the personification becomes a theophany, which is the appearance of God, Himself, to mankind. Another possible theophany is in Matthew 3:11 and Luke 3:16, which state that Jesus would baptize His followers in fire. This was later described as "tongues like as of fire" in Acts 2:3–4, and many scholars agree this was a literal manifestation of the Holy Spirit's presence. (Beale does not see a connection between the *fire* in the wormwood verses to a personification or theophany, and that's not where we are going with this; our purpose for pointing this out here is to express another layer from which *fire* cannot be limited to a literal interpretation only.) So already we see that something connected in the Word to fire might be subject to a deeper analysis, especially as it appears in the KJV translation 549 times in 506 verses.

Most pertinent to our study, however, are moments when *fire* represents a symbol of the wrath of God or a judgment (emphasis added in the following):

- God's "wrath" is "fire": "How long, LORD? Wilt thou hide thyself for ever? Shall *thy wrath burn like fire?*" (Ps. 89:46); "And I will pour out mine indignation upon thee, I will blow against thee in the *fire of my wrath*, and deliver thee into the hand of brutish men, and skilful to destroy" (Ezek. 21:31); "Therefore have I poured out mine indignation upon them; I have consumed them with the *fire of my wrath*: their own way have I recompensed upon their heads, saith the Lord GOD" (Ezek. 22:31); "For in my jealousy and in the *fire of my wrath* have I spoken, Surely in that day there shall be a great shaking in the land of Israel" (Ezek. 38:19).

- God's "rebuke" is "fire": "For, behold, the LORD will come with fire, and with his chariots like a whirlwind, to render his anger with fury, and *his rebuke with flames of fire*" (Isa. 66:15).

- God's "fury" is "fire": "He hath bent his bow like an enemy: he stood with his right hand as an adversary, and slew all that were pleasant to the eye in the tabernacle of the daughter of Zion: he poured out *his fury like fire*" (Lam. 2:4).

- God's "jealousy" is "fire": "Therefore wait ye upon me, saith the LORD, until the day that I rise up to the prey: for my determination is to gather the nations, that I may assemble the kingdoms, to pour upon them mine indignation, even all my fierce anger: for all the earth shall be devoured with *the fire of my jealousy*" (Zeph. 3:8).

There are more examples, but you get the idea. *Fire* is repetitiously used throughout the Word to represent the judgment of God upon the earth. With this in mind, one of the most powerful arguments for *fire* as a judgment of God *in the form of famine* is Ezekiel 5.

The prophecy regarding the *thirds* (a fraction that is becoming increasingly important by this point in our study) begins with friction between Israel and (ironically) the Babylonian armies, during the symbolic act of cutting Ezekiel's hair (vv. 1–4), which scholars believe stood for Israel's mourning (Jer. 45:3–5) and shame (2 Sam. 10:4–5). The figurative speech that sets up the entire chapter's description of judgment upon God's wayward people is established thus:

> [Ezekiel], take thee a sharp knife, take thee a barber's razor, and cause it to pass upon thine head and upon thy beard: then take thee balances to weigh, and divide the hair. Thou shalt burn with *fire* a third part in the midst of the city, when the days of the siege are fulfilled: and thou shalt take a third part, and smite about it with a knife: and a third part thou shalt scatter in the wind; and I will draw out a sword after them. Thou shalt also take thereof a few in number, and bind them in thy skirts. Then take of them again, and *cast them into the midst of the fire*, and *burn them in the fire*; for thereof shall *a fire come forth into all the house of Israel.*
> —EZEKIEL 5:1–4; EMPHASIS ADDED

It seems pretty clear by this point that fire is judgment, but obviously not a literal flame that engulfs and kills *all* the house of Israel. Looking a little way down the road in the same narrative, we read what this fire actually refers to:

> And I will do in thee that which I have not done, and whereunto I will not do any more the like, because of all thine abominations. Therefore *the fathers shall eat the sons in the*

midst of thee, and the sons shall eat their fathers [hunger to the point of cannibalism]....

A third part of thee shall die with the pestilence [disease], and with *famine* shall they be consumed in the midst of thee....

Moreover I will make thee waste, and a reproach among the nations that are round about thee, in the sight of all that pass by. So it shall be a reproach and a taunt, an instruction and an astonishment unto the nations that are round about thee, when I shall execute judgments in thee in anger and in fury and in furious rebukes. I the LORD have spoken it. When I shall send upon them *the evil arrows of famine*, which shall be for their destruction, and which I will send to destroy you: and I will *increase the famine upon you*, and will break your *staff of bread*: So will I send upon you *famine and evil beasts*, and they shall bereave thee: and pestilence and blood shall pass through thee [lack of food = weakness; weakness = vulnerability to disease]; and I will bring the sword upon thee. I the LORD have spoken it.

—EZEKIEL 5:9–17, EMPHASIS ADDED

Beale is not alone in seeing this, a judgment that God acknowledged to be unlike any before or after (Ezek. 5:9), as a clear and present indicator that fire as a judgment is likely to mean famine. The rest is a matter of showing the trumpet judgments of Revelation to be metaphorical.

First trumpet

With the trail linking the "fire" to "famine" here, Beale begins his analysis of the first trumpet blast by stepping back into Old Testament parallels, noting the similarity (yet again) of the thirds pattern. Zechariah 13:8–9 states: "And it shall come to pass, that in all the land, saith the LORD, two parts therein shall be cut off and die; but the third shall be left therein. And I will bring the third part *through the fire*, and will refine them as silver is refined, and will try them as gold is tried: they shall call on my name, and I will hear them: I will say, It is my people: and they shall say, The LORD is my God" (emphasis added). The antique writings recorded in the *Midrash Rabbah* document that this language, to the ancients, identified the one-third of the people who are *refined* through *fire* as Israel, and the *two parts* that are *cut off* and left to perish as the pagans.[16]

Using this as a guiding marker, let's look again at what Revelation 8:7 describes: "The first angel sounded, and there followed hail and *fire* mingled with blood, and they were cast upon the earth: and the third part of trees was burnt up, and all green grass was burnt up" (emphasis added). Beale points out that classic commentators, including John Gill, believe the *trees* in this verse to be heathen *kings*, as this follows the same language/terminology treatment from other Hebrew study texts, most specifically the first-century Targum.[17] The takeaway here, Beale asserts, is that the first trumpet judgment will fall upon both the righteous and the wicked, though the results will be different for each: The righteous will be *refined* and made stronger through trial, while the judgment will be *punishment* for the wicked, but no part of the population will completely avoid the effects of this trumpet.

As an interesting addition, consider the similar language used in Isaiah 10:16–20:

> Therefore shall the Lord, the Lord of hosts, send among his fat ones leanness; and under his glory he shall kindle a burning like the burning of a fire. And the light of Israel shall be for a fire, and his Holy One for a flame: and it shall burn and devour his thorns and his briers in one day; and shall consume the glory of his forest, and of his fruitful field, both soul and body: and they shall be as when a standard-bearer fainteth. And the rest of the trees of his forest shall be few,

that a child may write them. And it shall come to pass in that day, that the remnant of Israel, and such as are escaped of the house of Jacob, shall no more again stay upon him that smote them; but shall stay upon the LORD, the Holy One of Israel, in truth.

Second trumpet

A quick glimpse at mountain references throughout the rest of Revelation also shows that it can be a nation, kingdom, or people group:

- "And I looked, and, lo, a Lamb stood on the mount Sion [Zion; Israel; God's people], and with him an hundred forty and four thousand, having his Father's name written in their foreheads" (14:1).

- "And here is the mind which hath wisdom. The seven heads are seven mountains, on which the woman [Babylon] sitteth" (17:9).

- "And he carried me away in the spirit to a great and high mountain, and shewed me that great city, the holy Jerusalem, descending out of heaven from God" (21:10).

As such, Beale makes the obvious connection that the *great mountain* that was *burning with fire* in Revelation 8:8–9 represents a judgment upon a kingdom, and then he shows same-book context support of this strong theory by pointing to imagery of the same event several chapters later, in 18:20–21: "Rejoice over her [Babylon], thou heaven, and ye holy apostles and prophets; for God hath avenged you on [or pronounced judgment upon] her [Babylon]. And a mighty angel took up a stone like a great millstone, and cast it into the sea [here we have repeat imagery of something being hurled into the sea from space or from heaven, and by an angel, no doubt], saying, Thus with violence shall that great city Babylon be thrown down, and shall be found no more at all."

For those of you who may still be on the fence, Jeremiah 51:25 directly calls the kingdom of Babylon a *mountain* and then switches the verbiage to "a burned out mountain" or a "burnt mountain" representing the kingdom *after* judgment. This is, perhaps, too close to the second-trumpet verses of burning mountains being cast into the sea to write it off as a coincidence.

That's not to say that I, Tom Horn, believe Beale's second-trumpet theory to be the only possible interpretation. But if the playout of the apocalyptic events *is* literal (asteroids, etc.), I will find it at least poetic and majestic that God's nature is immutable *regardless* of whether His will is carried out literally, as the plagues of Egypt, or symbolically.

This, in addition to the earlier mentioned verses from Enoch that clearly wrap fallen angels and their judgment into "stars" and "mountains burning like fire" terminology, we already have a strong case for a metaphorical interpretation of the first two trumpets. Yet Beale, in his thorough thread, points at yet another imagery link to Babylon being cast into the Euphrates River (a freshwater source) and sinking into the waters (Jer. 51:63–64). It is at this point that the scholar acknowledges even his own *figurative* interpretation cannot irreversibly dismiss the possibility of *literal disasters*, but that "the burden of proof is on those who hold to a literal understanding *in addition to* a figurative perspective."[18] He adds that if the judgment *is* symbolic of famine, then it stands to reason that the destruction of the fish under the water represent a central source of food (which would have been true when Revelation was written and you couldn't drive to McDonald's), and the ships on the surface represent the vehicles that would transport food (i.e., "make the food attainable," possibly a link here to a severance of agricultural food trades that we so highly depend upon today).

Third trumpet

We now arrive to the trumpet judgment involving our old friend by the name of Wormwood. By now little remains to challenge how some scholars such as Beale and Heiser would see this "great star...burning" (Rev. 8:10–11) as a fallen angel and the ensuing judgment resulting in famine. Whether they are right or wrong, once their trail is followed from the onset to this moment, their *figurative-speech* and *historical/literary-context* theory holds a great deal of clout.

As has already been shown with Michael pages back, angels can, like mountains, also represent people groups or nations. Beale takes a moment to remind his readers that the plagues of Egypt involved direct judgment of Egyptian "gods" (more on this later). There appears to be herein a widespread connotation that water into blood, famine, and so on can be viewed to be God's smackdown of wicked entities—Beale refers to them as "legal agents representing sinful people."[19] The specific identity of this angel was listed in a quote by Beale earlier in this chapter, but now, after greater reflection, that quote bears repeating, as I anticipate it will land on readers' minds with fresh perspective: "The observation that the descent of the burning mountain in [Rev. 8:10] is parallel to the descent of the burning star in v 8 also indicates that the star should be identified as an angelic representative of an evil kingdom undergoing judgment. Here the judgment of Babylon's angel is in view, since v 8 concerns the judgment of Babylon the Great. The identification of the star as Babylon's representative angel becomes more convincing if v 10 is understood as alluding to Isa. 14:12–15."[20]

And speaking of fresh perspective, let's look at two of the other wormwood references elsewhere in Scripture: "Therefore thus saith the LORD of hosts, the God of Israel; Behold, I will feed them, even this people, with wormwood, and give them water of gall to drink" (Jer. 9:15). "Therefore thus saith the LORD of hosts concerning the prophets; Behold, I will feed them with wormwood, and make them drink the water of gall: for from the prophets of Jerusalem is profaneness gone forth into all the land" (Jer. 23:15).

What do these verses describe?

You might have guessed famine.

And you would be correct: "I will surely consume them, saith the LORD: there shall be no grapes on the vine, nor figs on the fig tree, and the leaf shall fade; and the things that I have given them shall pass away from them" (Jer. 8:13).

All of the famine judgment that we've discussed this far has been a result of one offense: idolatry. Beale drives it home: "So likewise in Rev. 8:11, Babylon, the prevailing world system, has influenced the earth-dwellers and some in the covenant community [God's people] to become idolatrous. And the consequence of such idolatrous pollution is judgment on both Babylon and those held under its sway."[21]

Same result either way?

I will conclude this section by pointing out that *all three* of the first trumpets, when interpreted as metaphoric representations, result in a continuous blast of famine against evil, idolatrous Babylon. They are linked, perhaps far more than we thought.

But here's the bottom line, in my opinion: Whether the Wormwood star itself is an angel or a space-body impact event, and whether something or someone is going to physically hit the waters of the earth or not, *might* we be facing the same thing anyway? Think about that for a moment...Angel *or* burning rock, poison water *or* famine—the result is still *pestilences and death*, as both the Old Testament and the New Testament qualify, and from all varying historical/literary context angles. Unless even the pestilence and death are metaphorical to some kind of political kingdom chess match (which, even then, would likely mean a destroyed economy, riots, chaos, and a number of other catastrophes leading to disease, hunger, and so on), we very well could all be talking about the same thing.

What does that scenario look like? Will we see the water turn red, the frogs jump to land, disease take our bodies, flies and gnats flocking to the smell of dead fish and livestock, and poison gases taking those who remain as a result of another Lake Nyos event triggered by eruptive phenomena or some equivalent? Or will nearly identical symptoms occur—hunger leading to increased vulnerability of disease, skin conditions, the smell of the death of livestock and people drawing winged pests, and so on—as a result of a Jeremiah- and Ezekiel-level famine upon evil Babylon?

I think the point made up to now is clear: Regardless of the vehicle used to execute judgment, there *will* be judgment, and it *will* affect every soul alive on the planet in that day. The wicked will be punished, and the righteous will be subject to a refining process that will nearly break them in what we can assume is severe suffering.

But there will be *no* hiding.

Chapter 6

AS IT WAS IN THE DAYS OF PHARAOH ... ETYMOLOGY OF *WORMWOOD*

A COMMON MISTAKE MANY make in approaching a prophetic topic is to skip straight past the relevant etymology (the study of the origins of words). Doing so holds the potential to lead to mass misinterpretation, and this has never been truer than with biblical Scripture. When these errors are made, the consequences can be devastating to our spiritual lives in many aspects.

Using an example from my previous work *Unearthing the Lost*

World of the Cloudeaters, I will illustrate why etymology *must* be the starting place for any discussion surrounding an obscure word:

> The origin of the word "obelisk" as far back as most etymology dictionaries can trace is the Greek *obelischos*, "small spit, obelisk, leg of a compass," which is diminutive of *obelos*, "pointed pillar, needle." This explanation, however, is limited by a modernized description of its shape alone, and only dates back to (at the very earliest) the first Greeks circa 1900–1600 BC....
>
> Etymology is the study of how a word was originally formed. The word "Internet," as one example, is a modern word. There is no ancient word belonging to what we today know as the World Wide Web because it didn't exist in the ancient world. When we say "Internet," people know we are describing "a vast computer network linking smaller computer networks worldwide...[through the use of] the same set of communications protocols," or more simply, that thing we log on to over the computer at night to check e-mail. So when we say "Internet," everyone knows what we're describing and what the word means in the same way many know that an obelisk is a "pointed pillar." However, the word "Internet" is a compound word made up of: 1) the prefix "inter-," meaning "'between,' 'among,' 'in the midst of,' 'mutually,' 'reciprocally,' 'together,' [and] 'during,'" and 2) "net(work)," meaning "a netlike combination of filaments, lines, veins, passages, or the like." Our culture has taken one prefix that means mutual, active cooperation between two or more sources and combined it with a word that describes a linking up of pathways. This is how etymology works.
>
> Fortunately, "Internet" is such a young word that we have no problem locating the etymological origins of it; unfortunately, "obelisk" is such an ancient word that many have a hard time locating the etymological origins of it. By the time "Internet" was a word, we had established worldwide communication, so the word is the same in many foreign languages,

despite slight variations of accent marks, dialect, pronunciation, and so on, and generally everyone is capable of understanding the prefix "inter-" alongside the noun "net" because they are words our modern world is familiar with. When "obelisk" was formed as a word, there was limited communication between one region and another, so one has to look back to that one specific region of the earliest of languages to figure out what "o" and "bel" and "isk" would mean to them when put together.

Herodotus, the Greek traveler, is largely attributed as the writer who named the object as he documented the cultures he visited and the structures he observed. As such, he would have had opportunity to look up at an edifice and call it by what it meant to the people of those lands but within his own language. In Greek, *o* is a prefix that frequently associates with "seed," *bel* means "lord" or "master," and *isk* is a suffix meaning "small." So the story goes that Herodotus was wandering about the civilizations of the ancient world, had prior familiarity with the concept that the structure was phallic to the original Egyptian carvers, and called it, essentially, "a smaller representation of the master's." If the *o* did mean to Herodotus (if he is, in fact, the coiner of the word) what it meant in other words with *o* as a prefix associating to "seed," then we arrive at: "a smaller representation of the master's seed [and by extension, his male reproductive organ]."

Which "master" would this refer to? That's where we get into the simpler etymology of the word *bel* to the first Greeks. "Belzebub" is Old English, based on Greek, which came from the original Hebrew for the Philistine god worshipped at Ekron, "*ba'al-z'bub*" (Hebrew; 2 Kings 1:2) and the Babylonian "sun god" or Baal, the "lord of the flies" when used by itself. Today, Christians have heard the name "Baalzebub" (more often "Beelzebub") and hear it as another name for Satan, but they very rarely understand that to the ancients, "Baal" was a prefix/title the Hebrews assigned to the foreign little-g "gods" that were worshiped by neighboring territories. (Consider as examples *Baal-berith* who was "the covenant lord" of the Shechemites and *Baal-peor* who was the "lord of the opening" over the Moabites and Midianites.) This renders our previous conclusion to read "a smaller representation of Baal's seed," or, more truncated, "the shaft of Baal."[1]

Because in *Unearthing the Lost World of the Cloudeaters* we did due diligence on the complicated etymological origins of the word *obelisk*, it went from meaning "pointed pillar" in our modern day all the way to "the shaft of Baal" to the original speakers. That is a *huge* leap in term definition!

The word *wormwood* is similar. Today it's most popularly known as a main herbal ingredient in the alcoholic beverage absinthe—the drink also commonly referred to as the "Green Fairy" due to Paris' late-1800s Bohemian hallucinogenic fairy icon as the liberator of artistic minds, as well as the drink's glowing green color. With that contemporary idea in mind, the "bitter water" of Revelation 8:10–11 sounds as if it would, at worst, make lightweight drinkers tipsy. Such a silly image demands that we do further digging. Let's start by eliminating what is *not* important to the discussion, since countless (and erroneous) fringe materials habitually tend to approach the name of the star from an anachronistic (chronologically inaccurate) perspective.

"Connections" to dismiss

First, the wormwood herb known to the biblical prophets and writers of Scripture would have been *Artemisia herba-alba* "(also called white wormwood), a small, heavily branched shrub with hairy, gray leaves…found even in dry, desolate areas,"[2] as opposed to *Artemisia absinthium*, and therefore probably has nothing to do with the traditional absinthe drink ingredient. Even if it could be shown that the herb of the Bible was *Artemisia absinthium*, the prophetic association to the Revelation star is that the waters will become "bitter" (poisoned in some way) and many will die from drinking it. It's fair to say that the alcoholic drink is completely irrelevant.

Second, although all expert etymologists openly acknowledge that all *Artemisia* plant classes were named after the Greek goddess Artemis (the goddess of the hunt and wildlife; also known as Diana and several other names) after the Hellenistic period, they have never truly agreed on how *worm* and *wood* were later put together in English to describe the herb. Sometimes a foreign word is respelled to match what it seems to be, resulting in a faulty and confusing history of origins (usually called "folk etymology" or "pseudo etymology"). As one example, *licorice* (a variant of the French word for "sweet root") was respelled *liquorice* in Britain and Ireland because of the mistaken assumption that the *licor* sound at the front had something to do with making *liquor* from the root. In the case of *wormwood*, many have assumed, understandably, that this is a compound description of the woody texture of the plant family's outer skin and the herb's historical connection to being used as a remedy for intestinal worms. Although this makes perfect sense, as far back as we can trace it, our English *wormwood*, from the Old English *wermōd*, compares with the German *wermut*, meaning vermouth (an herb-infused wine). So other than to assign a proper name to the destructive star, the English word *wormwood* holds no relevance to Revelation's star of destruction *or* to the herb.

Third, though it would be interesting to suggest that the Greek word for the plant family *Artemisia* connects our poisoning star Wormwood to an official pagan entity/deity (Artemis), there is no legitimate prophetic link here. Artemis is pictured in classical mythology to be the goddess of the wilderness and beasts, which relates to her role as goddess of the hunt, while just as much (if not more) ancient imagery depicts her as the many-breasted goddess of fertility, long life, sexual fulfillment, and protection during pregnancy and childbirth. The *Artemisia* herb was once thought to be a painkiller in childbirth. While there is no evidence remaining to indicate whether this was effective, in ancient times women on the brink of delivering a child would be praying to Artemis for protection while being given the herb *Artemisia* to ease that pain. Whereas this certainly may account for why the Greek name for *wormwood* involves this pagan deity, quite simply the Greeks renamed the herb after their pet goddess subsequent to wormwood's being a recognized medicinal plant as far back as Deuteronomy. Making any claim that the capitalized, proper-noun Wormwood star is the goddess Artemis (or that it is in any way connected with her) perpetuates an anachronistic, folk-etymology error.

Suffice it to say that the falling star of Revelation 8:10–11 is not related to absinthe, wood, worms, or the goddess Artemis. As I will now point out, the star also has nearly nothing to do with the herb.

Connections to keep

Before there was ever a "wormwood," there was an Old Testament Hebrew metonymy. A metonymy (from Greek *meta,* change, and *onoma,* name) replaces one name for another that closely associates with it, and the Bible is packed with occasions of this literary device. In the case of *wormwood,* translated from the Hebrew *la'anah,* not one reference in Scripture is ever talking about the herb. Although there is an obvious metaphorical connection to the herb's effects known to people of that time, *la'anah* (sometimes translated "gall"), according to *The Anchor Yale Bible Dictionary,* is "always used figuratively for bitterness and sorrow." This source goes on to explain:

> Deut 29:18 warns against the fruit of idolatry which is gall and wormwood (RSV: poisonous and bitter fruit). The prophet Amos describes perverted justice and righteousness as wormwood (5:7; 6:12). Jeremiah declares the judgment of God

against the people of Judah, saying, "Behold, I will feed this people with wormwood, and give them poisonous water to drink" (9:15; cf. 23:15). The author of Lamentations compares his distress over the destruction of Jerusalem to being filled with bitterness and wormwood (3:15, 19). In Proverbs the loose woman is portrayed as a deceiver whose lips drip honey, but who in reality "is bitter as wormwood" (5:3–4).[3]

The *Holman Illustrated Bible Dictionary* states that Old Testament prophets "pictured wormwood as the opposite of justice and righteousness."[4]

When you see the name of the herb—even when it's capitalized as a proper name once in Revelation 8:11—understand that you are reading a metonymy referring to a bitter *curse*.

The authoritative Logos Bible Software's *Concise Dictionary of the Words in the Greek Testament and the Hebrew Bible*, vol. 2, openly makes this connection, stating that *la'anah* is "from an unused root supposed to mean to *curse; wormwood* (regarded as *poisonous*, and therefore *accursed*)."[5]

Therefore, just as *obelisk* ultimately means shaft of Baal, wormwood in the biblical sense could be alternatively read: "And the name of the star is called Curse: and the third part of the waters became accursed; and many men died of the waters, because they were poisoned."

It's interesting to bring some more accurate perspectives forward with a little background digging, is it not?

THE CHRONOLOGY OF REVELATION

While it may sound odd, there are some scholars who assert that the events of the Book of Revelation may not necessarily be in chronological order. The confusion of timeline can emerge due to the fact that some say since the vision as given to John occurred within the spirit realm and held a linear quality, the events were written with intent to record *fully*, listing things in abbreviated form, then subsequently elaborating on the same happenings by repeating their occurrences in more detail.[6]

The assertion that there are parenthetical portions in the Book of Revelation is not a new idea. For many scholars and honest seekers of God's Word, the statement that there are portions of this prophetic work that deviate from the chronological timeline of events is not to infer in any way that these passages are not inspired, important, critical to the understanding of the whole, or that they should be disregarded or set aside. Contrariwise, from a cyclical, Hebraic standpoint events *could* roll out somewhat differently than has been traditionally preached in more recent decades. Whereas Western thinking tends to be very literal, demanding us to identify "[this] as absolutely [that]," a biblical mindset is better geared to ascertain how God has operated in the past, identify those familiar patterns, and safely assume that He will operate in the same fashion in the future, as it remains within His character and will to do so.

Pastor Richard A. Coombes, author of *America, the Babylon*, addressed the parenthetical nature of Revelation and offered a good overview of the intricacies included in the "things which will be." Coombes wrote that as the final book of the Bible opens, an angel appears to the apostle John and gives him an outline of three main categories into which everything that follows will drop.

1. The "things which you have seen" or "past events" pertain to Christ Jesus and the blessings incurred by paying attention and heeding the things in this vision.

2. The seven churches of Revelation and the evaluation given to/about them make up the "things which are."

3. Chapters 4 through 22:21 reveal "that which will be hereafter," and this carries us through "future events"

and the wrapping up of the current events in the earthly era timeline.[7]

Breaking it down further: Chapters 4 and 5 address events as they occur in heaven; 6–8:1 tackle the seven seals of judgment. According to Coombes, there is a switch to a different or parenthetical event in heaven shown in 7:1–17, revealing the sealing of the 144,000 witnesses, as well as worship by countless believers/ converts. Chapters 8–11 describe the seven trumpet judgments, which we have been focused upon in this work in the discussion of the star Wormwood. Between trumpets six and seven, Coombes explains that we are carried through a parenthetic description/story event regarding the mysterious "Little Book" and the two witnesses. Following this, Coombes cites three parenthetic events/symbols in a row from chapters 12 to 14: astronomy is shown in the description of the woman, child, stars, and dragon imagery, which correspond to chapters 6–11; two beasts target saints on earth and the entire world system in chapter 13; and in chapter 14 we see the Lamb with the men's choir of 144,000, an angelic call to the gospel, further bad news about Babylon, blessedness of the martyrs, and a sneak peak of Armageddon. This is the fifth parenthetical passage Coombes shows in Revelation. Chapter 15 introduces the "seven angels having the seven last plagues" and the singing martyrs before moving into chapter 16 with the *bowl* or *vial* judgments, which will be the last series of sequential judgment events. Also parenthetical in nature, chapter 17:1–9 gives an overview of the whore of Babylon, verses 10–11 offer a picture of the geopolitical progression influenced by Babylon, and verses 12–17 describe the Antichrist (the beast from chapter 13) rising up into a position of global control.[8]

Since the two beasts are introduced in chapter 13, Coombes poses an intriguing question: "So, if Revelation is in totally complete chronological order, why is it that we do not get a description of the rise of the Antichrist coming into power until midway through Chapter 17?"[9]

In any case, one can understand why some interpreters conclude that Revelation is not always chronological. There are portions of it that are placed in a very direct order (such as the first trumpet being followed by the second, which is trailed by the third, and so on), and it's far less likely that those sections are to be read in any chronology other than the literal arrangement given in John's vision. However, some maintain that to say with certainty that *every moment* of Revelation is sequential to the next may not be accurate. It is vital, however, that one searches diligently and prayerfully through Scripture in pursuit of the answer to this question, since all scriptural interpretation of prophecy will hinge upon an individual's clarity on this single point.

CHERNOBYL AND THE SECRETS OF FÁTIMA

Some readers may be familiar with other Wormwood or Revelation theories based on ancient prophecies from around the world, and they may be wondering at this point why this book hasn't addressed the almost-endless list of them. Quite frankly it's likely because these eschatological hypotheses are drawn from even weaker conjecture and presumptions than Sitchin's self-taught, zero-formal-education, ignoring-expert-research Sumerian cuneiform tablet translations regarding an Anunnaki-inhabited planet past Pluto on a trajectory toward Earth. They are most often pieced together by well-meaning folk who admittedly have enviable intelligence and imaginations but whose personal education and scholarly experience form conclusions that invariably disagree with historical, archaeological, anthropological/cultural (etc.) facts and research.

Most of these we don't plan to address at all because of the plethora of materials already in existence that show these theories to either be completely incorrect or entirely irrelevant (or both).

However, there are a couple of Wormwood theories that have gar-nered enough support in recent years that we will tackle them briefly.

May 13, 1917, marked the first of six visits that three children—Lúcia dos Santos and cousins Francisco and Jacinta Marto—claimed Mary, mother of Jesus, paid them. The sixth and final visit would occur in October of the same year. This Marian apparition has since been famously named Our Lady of Fátima. Since the events of 1917 the Marian visitations have been popularized and considered true by many within the Catholic Church due to the alleged con-nections between the prophecies of the apparition and World War II. However unrelated all that might sound to Wormwood (and for good reason), the apparition left the children with three secrets: the first is a vision of hell; the second, a warning about the conver-sion of Russia; and the third is an extremely complicated warning that has resulted in multiple interpretations and analyses (every-thing from a generic "pray hard, and you'll be saved" message to the bold assertion that the Church of Rome will eventually become the channel through which the Antichrist will rise to power).

Because people are apt to find connections where they will, there have been a lot of apocalyptic theories that conjoin all three of the secrets together with Nibiru, the Mayan Calendar, vague prophe-cies of Nostradamus, conspiracies of the Vatican in Rome, and even some restricted intel belonging to Adolf Hitler. By the end of the day some of these chain links are so far-fetched and wild that it would take a book at least the length of this one to dispute them all. Therefore, I will stick to just the most popularly referenced Fátima claims regarding Wormwood.

Our concern is with secret two, which states in part:

> You have seen hell where the souls [of] poor sinners go. To save them, God wishes to establish in the world devotion to My Immaculate Heart. If what I say to you is done, many

souls will be saved and there will be peace. The war [WWI] is going to end, but if people do not cease offending God, a worse one will break out during the pontificate of Pius XI [this was WWII, and it broke out just *barely* at the end of Pius XI's life and leadership in 1939]. When you see a night illumined by an unknown light, know that this is the great sign given you by God that He is about to punish the world for its crimes by means of war, famine and persecutions of the Church and of the Holy Father.

To prevent this, I shall come to ask for the consecration of Russia to my Immaculate Heart.... If my requests are heeded, Russia will be converted, and there will be peace; if not, she will spread her errors throughout the world, causing wars and persecutions of the Church. The good will be martyred; the Holy Father will have much to suffer; various nations will be annihilated. In the end, my Immaculate Heart will triumph. The Holy Father will consecrate Russia to me, and she will be converted and a period of peace will be granted to the world.[10]

On Saturday, April 26, 1986—after the number 4 reactor was experiencing some technical difficulties at the Chernobyl Nuclear Power Plant in Pripyat, Ukraine—a power surge ruptured the reactor, resulting in massive radioactive contamination that exploded into the atmosphere for nine straight days, spreading into parts of the USSR and Western Europe. Within days just under 120,000 people were evacuated from the area, growing to more than 300,000 in the coming years. Containment endeavors since the day of the accident to as recently as the 2017 Chernobyl New Safe Confinement project have continued, and the estimated completion year for cleanup is said to be 2065, making this the worst nuclear disaster in the history of the world.

Some say that exactly ten years to the day before the nuclear accident, another Marian apparition floated in a giant cloud above the sky, dropping dried sprigs of the wormwood herb to the ground in a warning of the upcoming explosion a decade later. (The lack of enough credible eyewitness accounts for an event so major—paired with almost no media coverage of this occurrence ever—places this story high on the probably-never-happened list.)

If secret two from Our Lady of Fátima was correct, then Russia would be in for a major humbling if it did not consecrate. If Chernobyl was *that* great humbling event, then certain bizarre language from this enigmatic foretelling would make sense, such as the "night illumined by an unknown light" (which could be interpreted to be either the Marian wormwood-cloud warning or poetic reference to the glow of nuclear radiation).

Now for the glaring link: the word Chernobyl is Ukrainian for wormwood.

Could the nuclear radiation outbreak of 1986 (ironically the same year as the Lake Nyos tragedy) have been the fulfilment of the Revelation prophecy?

There are abundant major issues with this theory (*apart* from the fact that it's a historicist approach, placing a behemoth-level, world-ending futuristic and prophetic event of Revelation in the past), but I will only outline five:

1. In direct response to secret two, Pope St. John Paul II, on March 25, 1984—*almost two years before the Chernobyl accident*—consecrated Russia to the Immaculate Heart of Mary. Many curious speculators have put forth this question: "If Russia *was* consecrated in 1984, then why wouldn't it be spared from the devastation of Chernobyl, if indeed Chernobyl *was* the prophesied disaster?" The answer comes from all corners and with more varying opinions than I can count, but the gist boils down to: "Well, the consecration of Pope St. John Paul II didn't satisfy what Our Lady wanted." However, according to

the Vatican website, Sister Lúcia *herself* in a letter dated November 8 of 1989, acknowledged that this consecration ceremony performed in 1984 *did* satisfy what the Fátima apparition commanded: "Yes," Sister Lúcia wrote, "it has been done just as Our Lady asked, on 25 March 1984."[11]

2. Secret two made it clear that we were facing *enormous* cataclysm for Russia's refusal to consecrate, including verbiage such as "war, famine and persecutions of the Church and of the Holy Father," "[Russia] will spread her errors throughout the world," "The good will be martyred," "various nations will be annihilated," and so on. Whereas this kind of judgment could perhaps more easily be linked to the events of WWII, the apocalyptic implication of the secret two prophecy is a vision of far greater international annihilation than the aftermath of Chernobyl has been, which brings me to my next point.

3. The casualties of the Chernobyl event cannot possibly be accurately calculated as a result of too many variables. In fact, depending on the report one reads today, the death toll is as little as four thousand or as many as two hundred thousand, all estimates from different health, science, or environmentalist organizations.[12] And despite speculation of the radiation poisoning done to the surrounding freshwater sources—primarily that of the Dnieper River—not one dependable calculation has ever been made that would prove one-third of the Earth's total freshwater sources were affected or ever would be. One would think that an event as devastating as the Wormwood prophecy of Revelation would be a little

less ambiguous and that the effects would be at least a *little* closer to the numbers prophesied.

4. Linking secret two of the Fátima apparition relies on dismissing most reliable scholars' educated verdicts on the subject. Up to this point in the book we have visited the possibility of a literal (space-body, asteroid, comet, eruptive phenomenon, etc.) water-poisoning event, as well as the metaphorical (angel, famine) event—but *both* of these views are on the wording of the Book of Revelation in regard to the entire world population at the time of the Antichrist's one-world-government reign, not an obscure, pre-WWII prophecy about the consecration of Russia. At best we have to willingly ignore a lot of scripture, most historical and literary context *behind* that scripture, and countless rules and principles of proper biblical interpretation and exegesis to arrive at the conclusion that Wormwood *was* the Chernobyl tragedy.

5. And finally, to put what I believe to be the final nail in the Chernobyl coffin: Yes, *chernobyl* is Ukrainian for *wormwood*; on paper that looks as if it means something. However, Chernobyl is also the city nearby the power plant of the same name, both dubbed such because the herb wormwood (which our etymology study showed to be unrelated) grows in that area. If wild roses had coincidentally grown there instead, the area of the nuclear plant explosion may have been called Roses or Troyandy (the Ukrainian word for *roses*), and the entire Wormwood connection would fall flat. That says something about

the weakness of the connection when a mere name crumbles the whole prophetic canvas.

In my humble opinion it's a bit of a stretch to take this prophecy by a Marian apparition more seriously than the Book of Revelation, and I believe the work of scholars such as Heiser, Beale, and numerous others makes it clear that Chernobyl and the consecration of Russia are unrelated.

Hopi Blue/Red Kachina prophecy

There is a common thread online that links an allegedly ancient Hopi Native American prophecy with Nibiru, Wormwood, the apocalypse, various verses from Revelation, the Hale-Bopp comet, and occasionally far-out predictions by Nostradamus.

The Blue Star Kachina, a dancing kachina spirit, will appear in the skies, signifying the birth of the new world. The emergence of this entity is the last sign before the Red Star Kachina turns the universe red and essentially destroys Earth as we know it during an event called the Day of Purification.

First, the evidence points to the idea that this prophecy is as recent as the twentieth century, so there may not be anything ancient about it.

Second, this Hopi prophecy can be looked at one of two ways for our purposes here: a) one tribe of Native Americans have their own culture's early reference to personified stars from space assisting in the destruction of Earth, and it is completely unrelated to Nibiru, or b) many websites are correct in their claim that the Hopi were in fact referring to the one-and-the-same planet Nibiru when they wrote about the Red and Blue Kachina stars. If the former, then I'm not surprised; it appears that *many* ancient cultures have similar apocalyptic prophecies, but ancient apocalyptic predictions from across the globe are not what this book is about, and this particular issue seems to be unrelated to the question of Wormwood. However, diving deeper into our study of ancient culture does render a different, *unexpected* angle, which may just have *everything* to do with Wormwood.

An Exodus Connection?

Some Word-weathered folks may have already made the following connections, but for as many times as I have been reminded, it still amazes me how recognizable God's fingerprints are from one historical era of judgment to another: trumpet one (hail, fire) had a predecessor in the seventh plague of Egypt (Exod. 9:22–25); trumpets two and three (water into blood, contamination of water supply) hold eerie similarities to the first plague (Exod. 7:20–25); trumpet four (darkness) is akin to the ninth plague (Exod. 10:21–23); and trumpet five (locusts) parallels the pests of the eighth plague (Exod. 10:12–15).

Not only do the plagues of Moses' day have value to us now as a way to further comprehend God's character, personality, and reactions to paganism and government leaders who persecute His people (as the pharaoh did and as the Antichrist will in the end times); from the damage these plagues had on Egypt, we can also gain an archetypal drawing board upon which we can imagine and calculate future Egyptian-like plagues visiting Earth during Wormwood's devastation in the great tribulation involving judgments against modern idolaters.

That readers may understand how God did (and will again) judge false gods and those who idolize them (see Revelation 9:20), allow me to excerpt the following summary from my now out-of-print book *The Gods Who Walk Among Us*, illustrating specifically which entities and people were condemned during the Exodus and what lessons this may convey for the days surrounding Wormwood.

The Nile plague

"Take thy rod, and stretch out thine hand upon the waters of Egypt...that they may become blood..."

Why would Yahweh turn the Nile into blood? Because the Nile was worshiped as the single most important element needed for the ongoing success of the culture, economy, and paganism of the Egyptian people. The annual flooding of the Nile brought new life and sustenance to over 1,000 miles of Egyptian-dominated settlements, and the watery event was perceived by the Egyptians as the best evidence that the gods of the Nile were pleased. When the Hebrew God challenged the welfare and divinity of the Nile River, He was striking a blow at the core of the Egyptian's faith and pantheon.

First, the waters of the Nile were esteemed sacred. Blood, on the other hand, was considered an abhorrence to the Egyptian people. The Nile River was supposedly protected from the contamination of human blood and other such impurities by the fearsome ram-headed god, Khnum, who consorted with Sati—the goddess of Elephantine—as the dispenser and protector of the cool waters. Second, the Nile was "possessed" by the spirit of Hapi, the son of Horus, who was often depicted as a corpulent man with the breasts of a female (representing the abundance and succoring of the Nile) and was honored as the god who, through using the silt and waters of the Nile, provided the abundant fertility of the land of Egypt. At other times Hapi was depicted as a mummified man with the head of a baboon, a portrayal in which he was considered the guardian of the lungs of the deceased and the Nile-servant of Osiris. Keeping Osiris happy was important to the welfare of the Nile, because the origin of the Nile was not known in ancient times (the central African location was not discovered until 1862) and the Nile's origin was considered by the Egyptians to be the spiritual bloodstream, or divine "life flow," of the netherworld Osiris. Turning the Nile into blood was thus, in part, a mockery of the Osiris blood-myth by the

Hebrew God, Yahweh. Fourth, the fish of the Nile were considered sacred and were supposedly protected by two powerful goddesses—Hathor, the goddess of the sky and the queen of heaven (who protected the chromis, or "small fish"), and Neith, the very ancient goddess of war who protected the lates (large fish), which were also considered to be her children. Neith was a powerful Egyptian deity, the sister of Isis, and the protectress of Duamutef—the god who watched over the inner stomach of the dead. More important, she was the mother of the Nile-god Sobek, an evil god with the head of a crocodile, to whom pharaoh may have "offered" the Hebrew male children when he commanded the midwives to throw them into the Nile. (Ex. 1:22)

Another Nile crocodile god—Apepi—was the arch rival of the sun god Ra, and may have been one of the "serpents" who appeared before Pharaoh and Moses in Exodus 7:10–12. We read, "And...Aaron cast down his rod before Pharaoh...and it became a serpent (tanniym, dragon or crocodile)...the magicians of Egypt, they also did in like manner with their enchantments...and they became serpents [crocodiles—Sobek and Apepi?]: but Aaron's rod swallowed up their rods." Since the word tanniym is not translated "serpent" anywhere else in Scripture, Dr. Jones [my research partner when writing *The Gods*] believes, as do other scholars, that tanniym should be interpreted as dragons or "crocodiles" in the Book of Exodus, as it was thus translated throughout the Books of Isaiah and Ezekiel. Either way, by turning the Nile River into blood, no less than nine deities were judged by the Hebrew God and found to be inferior and under His authority; the Nile River, Khnum, Sati, Hapi, Osiris, Hathor, Neith, Sobek, and Apepi. Through the first plague Yahweh confirmed that He alone is the supplier of every human need, and the true Judge of the after life and only Sovereign of destiny.

The frogs plague

"...behold, I will smite all thy borders with frogs: And the river shall bring forth frogs abundantly..."

When I was a young Christian I had an interesting experience during a time of intercessory prayer. I was fasting and praying for the salvation of a member of my wife's family when suddenly the image of a frog appeared before my mind's eye. The vision startled me because it was unexpected and powerful. No matter how I tried, I could not shake the uncanny feeling that a "frog" was resisting my prayer. It had the appearance of a typical river frog, but it stared at me as if to warn me that I had wandered into its "territory" and that it was fully intending to defend its position within the life of the person for whom I was praying. After a while it became obvious that whatever or whoever the frog creature was, it was not going away, and so I rebuked it "in the name of Jesus," and it immediately vanished. Some time later I was amazed to discover that certain demons can appear in the images of frogs. In the Book of Revelation we read: "And I saw three unclean spirits like frogs come out of the mouth of the dragon, and out of the mouth of the beast, and out of the mouth of the false prophet" (Rev. 16:13). My experience had been genuine, and, over time, helped me to understand that the Hebrew depiction of frogs as unclean animals was perhaps based on an ancient and spiritual revelation from Yahweh.

To the ancient Egyptians, however, frogs were sacred animals, and, ultimately, the infants of the frog-goddess Heka. Heka played an important role in the development of infants, including humans, beginning at the embryonic stage and continuing until childbirth. She was thus an important patroness of midwives and a powerful goddess of fertility. As the wife of Khnum, she assisted in the original creation of mankind and was closely associated with Hapi, who held the divine frog in his hands as the waters of nourishment flowed from her mouth.

When the krur (frogs) increased along the banks of the river during the annual receding of the Nile, it was perceived by the Egyptians as a good Heka omen. It's easy to see how the Plague of the frogs was a substantial embarrassment to the Egyptians—to have the frog-goddess babies so multiplied that one could not walk upon the ground or move within the house without squashing the divine creatures beneath their feet. Pharaoh could not order the Hebrew slaves to destroy and haul the frogs away, as it was a capital offence to kill a frog in Egypt! How powerful the Hebrew Creator-God must have appeared compared to the stupidity and stench of the creator-frog goddess, as her infants lay rotting in massive filthy heaps, covering nearly every square inch of Pharaoh's Egyptian empire. Through the plague of the frogs, the mystical power of Heka was reduced to nothing more than a greasy pavement crushed beneath the feet of the sorrowful Egyptians.

The lice plague

"...Stretch out thy rod, and smite the dust of the land, that it may become lice throughout all the land of Egypt..."

Four distinct areas stand out in the plague of the lice: First, the priests of Egypt were immaculate regarding their purifications. Discovering a single louse would have rendered an Egyptian priest as unclean, and, as such, incapable of ministering in the temple. The shutdown of the priestly ministry would have been no small matter, as the priesthood numbered in the thousands of men who maintained a strict regimen of daily ministering, bathing, shaving, and of sacred purification. Such priests exercised great influence over the common Egyptians, and were considered the uppermost servants of the gods—the "consecrated ones" who carried out the required daily ceremonies of the hundreds of Egyptian temples deemed necessary for the ongoing functionality of the local community. The religious duties of the priests included two main categories: 1) carrying about the little shrines or oracles (small replica temples containing statues of the gods) which were made available to the common people (those who could not enter beyond the veil); and, 2) performing the mystery rituals in the inner sanctums or "holy of holies" of the temples. The difference between the two priestly categories was that the portable gods were publicly available to nod their heads and speak (it's been suggested that the priests spoke for the idols while moving their mouths with a string) while the mystery functions of the priesthood were highly secretive and included the important creation rituals conducted in the inner sanctums of the main temples, like those of Amun-Ra at Karnak, where a priestess known as the "hand of god" performed ritual masturbations on the priests as a form of imitative magic (referring to the Amun masturbation/creation myth). This practice was considered necessary for the ongoing balance of nature, the annual flooding of the Nile, and regulating the seasons.

Moses was a man "learned in [such] wisdom of the Egyptians…" (Acts 7:22). As such, he was aware of, and may have been trained in, the mysteries of the Egyptian priesthood. It's even possible that Moses served as an Egyptian priest. The name Moses means to be "drawn out of" or "born of" and was usually associated with a priestly Egyptian deity, i.e., Thothmoses (born of Thoth), Amenmosis (born of Amen), or Rameses (born of Ra). The slight variations of the spelling of Moses (mosis, meses, etc.) did not change the priestly Egyptian meaning. This has caused some scholars to conclude that the Hebrew Moses may have been named after a Nile deity by the Pharaoh's daughter (Ex. 2:10), and that he served as an Egyptian priest who later dropped his Nile deity-name reference upon encountering the omnipotent Yahweh—the God of his fathers. Whether or not that's true, Moses was raised in the Pharaoh's court, thus he had a special understanding of the far-reaching ramifications of the plague of the lice. Moses understood that, when every micro-particle of dust began to crawl upon the Egyptians, the priesthood was ceremonially unclean, and thus immobilized. The masturbation mysteries of Karnak could not be performed! The portable gods could not walk and talk! The seasons could not bring forth their blessings! While this kind of reasoning may seem simplistic, such imitative magic, as performed by the Egyptian priesthood, was central to the Egyptian way of life and was considered of the highest importance.

The second point of interest concerning the plague of the lice involves the fact that Pharaoh was supposedly the incarnation of Horus and the son of the Sun god Ra. He was thus god incarnate. The dust of Egypt was therefore holy ground. To say the least, it was a serious slap in the Pharaoh's face for the Hebrew God to transform the sacred dust of Ra into lice.

The third notable point concerning the plague of the lice is that the magicians (priests) of Egypt could not duplicate the miracle, as they had duplicated the first two plagues. The Hebrew God was perhaps illustrating that He alone has the ability to create life out of the dust of the earth. Even the magicians testified, "This is the finger of God" (Ex. 8:19).

Fourthly, Geb (earth) was the god who protected the soil, while Seth was, among other things, the angry god of the desert sand. In the Osiris myth it was Seth who raged against the other gods in his bid to become the greatest among the Egyptian pantheon. In the plague of the lice, Yahweh was perhaps mocking the Egyptian religion by causing the dust god Seth to literally struggle against Geb, Ra, and Osiris, while leaving the Egyptians to suffer as collateral casualties.

The flies plague
"...and there came a grievous swarm of flies...and...the land was corrupted by reason of the swarm of flies..."

The beetle was known in Egypt as a fly. The scarab beetle was the sacred emblem of the sun god Ra and was the symbol of eternal life. But the flies of the fourth plague were most likely a blood-sucking breed that spread blindness and disease among the populace that lived along the Nile. Whereas such flies were generally disliked by the Egyptians, they were, nevertheless, revered as the servants (demons?) of Vatchit—the Egyptian "lord of the flies." In this context it's possible that the Hebrew God was administering a threefold judgement: First, of the Egyptians for their veneration of the fly-deities; second, of the sun god Ra—the Egyptian almighty creator; and third, of Vatchit himself, the Egyptian equivalent of Baalzebub (Beelzebub), the very ancient god who, according to various eastern religions, was the Evil god and "lord of the flies."

The name Baalzebub originally derived from two different words; Baal (lord, master), and zebub (of flies). While the original meaning is unclear and may have referred to a certain priestly interpretation of the flight path of flies as an oracular communication between Baal and his followers, some have pointed out that flies are adjoined to decaying bodies, and thus Baal-zebub may have been a kind of Baal-Osiris; a lord-demon of the human corpse.

According to the Grimorium Verum and the Grand Grimoire (18th century textbooks on magic) Baalzebub manifested himself in the image of a huge fly whenever he was summoned by the sorcerer. Whether Baalzebub, like Vatchit, commanded the flies to do his bidding, or delivered from their nuisance, is unclear. Additionally, since the title "Baal" referred to any lord, deity, or human master, there were many gods of antiquity known as a Baal; i.e. Baal-berith (lord of the covenant), Baal-Gad (lord of the fortune), Baal-hazor (lord of the village), and so on. That some Baals were worshiped by the Egyptians is known from the titles of certain Egyptian provinces; i.e. Baal-ze-phon ("Baal-of-the-North" or "Hidden Place") of Exodus 14:2. In times of great distress it was usually a Baal that was called upon for help, and people who sought material prosperity believed their lives could be improved by offering their firstborn child as a sacrifice to the deity. The

Greek author Kleitarchos recorded the dastardly process of sacrificing infants to Baal three hundred years before Christ:

'Out of reverence for Kronos [Baal], the Phoeniciens, and especially the Carthaginians, whenever they seek to obtain some great favor, vow one of their children, burning it as a sacrifice to the deity, if they are especially eager to gain success. There stands in their midst a bronze statue of Kronos, its hands extended over a bronze brazier, the flames of which engulf the child. When the flames fall on the body, the limbs contract and the open mouth seems almost to be laughing [such areas of child sacrifice were often called "the place of laughing"], until the contracted body slips quietly into the brazier.'

The sacrifice of babies to Baal was widespread in antiquity and was practiced by the children of Israel under the reign of King Ahab and Queen Jezebel. A recent archaeological find illustrated how far-reaching such offerings were. It unearthed the remains of over 20,000 infants that had been sacrificed to a single Baal. Ahaziah, King of Israel, may have authorized such a child sacrifice when he sent messengers to the Philistine city of Ekron to inquire of the fly-god as to whether he (the king) would recover from his illness. Yahweh intervened and instructed Elijah to prophesy to the King; "Thus saith the Lord, Forasmuch as thou hast sent messengers to inquire of Baalzebub the god of Ekron, is it not because there is no God in Israel to inquire of his word? therefore thou shalt not come down off that bed on which thou art gone up, but shall surely die" (2 Kings 1:16).

The Hebrews acknowledged Baalzebub as Satan's highest power and often referred to him as Beelzeboul, "lord of the height," a classification in which he was depicted as the dark atmospheric god who controlled the kosmos, or circumambient "air." Baalzebub eventually developed into a demon-god of such evil reputation that he became known as "the prince of devils" (Mt. 12:24). Milton referred to Baalzebub in *Paradise Lost* as Satan's chief lieutenant, and, in the litanies of the witches' Sabbath, Baalzebub is ranked, together with Lucifer and Leviathan, as an equal member of the supreme trinity of evil. The mocking of Vatchit as the Egyptian equivalent of Baalzebub, and thus as the ultimate manifestation of evil, may have been what the Hebrew God had in mind during the plague of the flies.

The murrain plague

"...Behold, the hand of the Lord is upon thy cattle which is in the field...there shall be a very grievous murrain..."

The sacred Apis bull of Egypt was a perfect example of [doctrinal plagiarism of the original revelation from God to Adam] in that the Apis bull was a demonization of the life of Jesus, especially of the protoevangelium—the biblical promise of an immaculately born Son of God, who would also be God in flesh. (See Gen. 3:15; Is. 7:14) Cattle, and especially the Apis bull, were sacred to the Egyptians. But the Apis bull (also known as Serapis or Osorapis) was special in that it was supposedly born of a miraculous conception when, every twenty-five years, divine moonlight (or lightning) struck a cow and it conceived. The Apis bull was thus considered to be the incarnation of god on earth. During its life span the Apis was worshipped as both the son, and incarnation, of Ptah—the Universal Architect god. As Ptah incarnate, the Apis embodied the Egyptian logos god who created, according to a later version of the Egyptian creation myth (the Memphite cosmology), all of creation by the authority of his spoken word. In death the Apis supposedly experienced a "resurrection" with Osiris, and thus the Apis bull, identified with Osiris, was a remarkable parallel of the Christian Messiah.

For practical reasons the Apis bull was, for the most part,

kept in seclusion. The Egyptian priests cared for the sacred animal and worked with a team of doctors and nutritionists in maintaining the bull's health. At the end of the twenty-five year cycle, a new bull was chosen with the pomp and ceremony of royalty. A celebration followed the selection and, for a period of forty days thereafter, the Egyptian women raised their dresses and exposed themselves to the bull. Such exposure was thought to capture the fertility energies of the Apis, and to excite the life-giving waters of Osiris. It was also believed that a special generational blessing came upon the exposed women's offspring. At the end of the forty days of "exposure", the new bull was removed to the Apis temple in Memphis where it was kept in a special sanctuary. It was thereafter publicly displayed during special occasions only. We find the Apis sanctuary mentioned in *The Geography of Strabo* (63 B.C.–A.D. 26):

> 'Memphis itself, the royal residence of the Aegyptians [Egyptians], is also near Babylon; for the distance to it from the Delta is only three schoeni. It contains temples, one of which is that of Apis, who is the same as Osiris; it is here that the bull Apis is kept in a kind of sanctuary, being regarded, as I have said, as god...'

Once the new Apis was inaugurated, the old bull was drowned, mummified, mourned, and placed into a huge sarcophagus. The burial rites of the passing bull were so revered and costly that they were paralleled in Egypt only by those of the Pharaoh. The comprehensive nature of such Apis burial rituals was illustrated in 1851 when 60 Apis sarcophagi of red and black granite weighing more than 60 tons each were discovered in Saqquara just west of Memphis where the Apis temple stood. I once took notes on, and photographed, the mummified head of one such bull. It was obvious from the detail and craftsmanship that great reverence was given to the animal during the mummification process. Especially impressive were the elaborate glass eyes which had been placed into the eye sockets, and the golden sun-disk of Ra that rested between the horns. Such golden discs were similar to the moon-discs worn by other members of the divine bovine family, including those donned by Hathor, the cow-goddess-mother of the sun god, Ra.

While other religions have practiced similar venerations of cattle—most notably Hinduism and in India where the humped Zebu cow continues to be worshipped as the representative of Aditi, the "sinless cow"—nowhere was the deification of such animals more noteworthy than in Egypt. The ancient Egyptians considered all cattle to be sacred sources of generative power, and the cults of Apis and of Hathor thus set the standards of eastern myth and ritual. This fact has caused the plague of the deadly murrain to be considered an especially effective grievance, as, in a single move, it repudiated the six most important aspects of the Apis cult: 1) it devastated the protected livestock of the Egyptians including the vast herds of Pharaoh; 2) it illustrated God's unlimited power when, miraculously, none of the Hebrew cattle died; 3) it humiliated the Universal Architect god, Ptah, and exposed him as a helpless demon; 4) it destroyed the dominion of the sacred Apis and Mnevis bulls of Heliopolis; 5) it judged the goddess Hathor, and the god Osiris, and found them to be inferior; and 6) it nullified the generational blessings of Apis-Osiris (Serapis).

Amazingly, after all of this evidence, the foolish heart of Pharaoh was hardened against the Hebrew God.

The boils plague

"...And it shall become small dust in all the land of Egypt, and shall be a boil breaking forth with blains upon man, and upon beast, throughout all the land of Egypt..."

During the third Egyptian Dynasty and at least 1,000

years before the Exodus, a man named Imhotep served as the vizier of the Pharaoh Zoser. Imhotep was an engineering genius and built the first-known massive stone structures, including the great Step-Pyramid (still standing) at Saqquara. From history we learn that Imhotep's well-founded distinction as a builder was surpassed only by his talent as a skilled magician and healer. When the Egyptians suffered under a seven-year famine which occurred during the reign of Zoser, the king appealed to Imhotep, who in turn consulted the sacred books. After several days Imhotep emerged from isolation and announced to the king "the hidden wonders, the way to which had been shown to no king for unimaginable ages." Zoser, impressed with Imhotep's discernment, obeyed the divinations. Simultaneously, Egypt withdrew from the famine and Imhotep was decreed the chief Kheri-heb priest ("son of Ptah") of Egypt.

But the popularity of Imhotep's life eventually gave way to the fame that followed his death, as later he was elevated, deified, and transformed into a healing god. By the time of the reign of the Pharaoh Menkaure (BC 2600), temples throughout Egypt were dedicated to the god Imhotep. Such temples contained incubation or "sleeping" chambers used in the convalescence of the sick and the mentally diseased, and the same became acknowledged as the most potent healing alchemies of Egypt. The incubation-temple of Imhotep at Memphis, for instance, proved to be so popular that the Greeks identified Imhotep with Asclepius, the Greek god of healing, and affirmed his divine membership within the powerful Egyptian "trinity" composed of Imhotep, Ptah, and Sekhmet the lion-headed goddess.

It's said that Imhotep convinced the Egyptians that premature forms of sickness and disease could be ultimately avoided if the proper aspects of healing-magic were carefully employed. The magicians of Imhotep used the magic crystals and incantations of Isis to call upon Sekhmet—the goddess-sovereign of epidemics and diseases—to work with the positive energies of Serapis in the administration of the healing needs of the Egyptians. Such rituals were often accompanied with burnt offerings, and the ashes of the same were sprinkled into the air as a health-blessing for the Egyptians. At other times the diagnosis called for an extended stay in the temple of Serapis where the sick or injured person was placed under the mystical spell of the katoche. Such katoche supposedly provided the internal coercion of the god and ultimately led to the proper diagnosis, and divine assimilation, of the transmissible and healing energies of the god. The katoche, affiliated with Imhotep's sleep-wizardry, was linked to the mystical crystals of Isis. These, in turn, were joined with Sekhmet's administration of the overall life-giving energies of Ptah and Osiris. Combined, they provided the Egyptian magicians with the powerful and esoteric tools necessary for the overall health (?) of the people. Such magic was indeed powerful, and the fame of such men and magic (Jannes and Jambres) continued up until the times of the New Testament. (2 Tim. 3:8)

When the Hebrew God attacked the divine health of the Egyptians by placing a filthy, eruptive disease of boils upon the population, He accomplished what no other surrounding power had attempted to do: 1) He sent the respected Egyptian magicians fleeing powerless before Moses—unclean and unable to perform their priestly duties; 2) He illustrated the inferiority of the Egyptian high gods—Ptah and Osiris— and denounced them as helpless demons; 3) He judged the lion-headed goddess Sekhmet and demonstrated her impotence at regulating diseases; 4) He altered the ritual of "casting ashes" and made the ashes a cursing instead of a blessing; and 5) He mocked the temples of Imhotep and Serapis, and thereby notified the surrounding nations that neither crystals, nor psychic dreams, nor positive energies, nor yet coercions of men and their gods, can defy the incontestable will of Yahweh.

The hail plague

"...Behold, tomorrow about this time I will cause it to rain a very grievous hail, such as hath not been in Egypt since the foundation thereof even until now..."

The goddess Nut was the Egyptian protectress of the sky and weather, and was depicted in Egyptian art as a woman arched over the earth, with the stars above her back and the earth (her brother Geb) beneath her belly. She was the consort of Osiris—the "blesser" of crops and fertility—and was cherished as the caring mother "sky-goddess" by the agricultural people of the Fertile Crescent. According to myth, Nut elevated herself each morning upon her fingers and toes and thereby provided an expanse between herself and Geb/earth. The spherical covering generated by Nut's towering action allowed the sun god Amun-Ra to coat the earth with light, and the warmth of the new day was received as a blessing of the goddess. At night, when Nut lay down, the expanse closed anew and darkness covered the earth. To the Egyptians, this was the natural order of things. But when violent storms erupted and the daytime skies were darkened, the same was perceived as a disturbance in the original cosmic scheme.

Nut was displeased with such nonconforming weather, and, at times, the skies grew red with the blood of her wounds (other myths define the red skies as Nut's menstrual period) as she struggled against the storm to restore the cosmic rhythm. For the sake of her people, the Egyptians, Nut bravely fought to maintain the essential universal cycle. Both men and gods depended on the cycle of Nut. Amun-Ra needed her expanse to visit the earth each day. Seth needed the same to dry the desert sand. Osiris needed Nut's meteorological blessings to sustain the agriculture; and Pharaoh desired the sanctions of Nut for two essential reasons: First, she controlled the atmospheric conditions surrounding the Pharaoh's Egyptian empire, and second she conquered the fierce storms that could herald the death of a king. For these and other reasons, Nut was particularly important within Egyptian devotions.

When the Hebrew God sent a storm of hail and fire "such as there was none like it in all the land of Egypt since it became a nation" (Ex. 9:24), He was repudiating the combined efforts of Nut, Geb, Amun-Ra, Osiris, and Pharaoh, to control the atmospheric conditions that befell the land of Egypt. A similar storm of fire mingled with hail is predicted to hit the earth again during the Great Tribulation. We read, "The first angel sounded, and there followed hail and fire mingled with blood, and they were cast upon the earth..." (Rev. 8:7). Just as Pharaoh rejected Yahweh, embraced pagan idols, and hardened his foolish heart, modern men seem destined to repeat the same mistakes. We find "...the rest of the men which were not killed by these plagues yet repented not of the works of their hands, that they should not worship devils, and idols of gold, and silver, and brass, and stone, and of wood" (Rev. 9:20). Such verses indicate a latter-day revival of idolatry, and provide the impetus for the last two chapters of this book in which we discuss the prophetic and extensive aspects of modern paganism.

The locusts plague

"...behold, tomorrow will I bring the locusts into thy coast..."

The locust plague was an awesome spectacle and was one of the most feared pestilences of the ancient world. Such invasions darken the sky and ingest every green thing. An average locust can consume its weight in food daily, and can quickly amass an army of insects numbering in the hundreds of millions per square mile. In 1927, a few African locusts were spotted near a river in Timbuktu. Within three years the whole of west Africa was besieged by the creatures. Eventually, locusts covered an area more than 2,000 miles wide—extending from Ethiopia and the Belgian Congo to the luxuriant farm lands of Angola. Finally, 14 years after the plague began, 5,000,000 square miles of Africa (an area twice the size of the United States!) had been destroyed by the locusts.

In ancient times the idea of such a calamity brought instant terror to the disposition of the vegetation-dependent Egyptians. To avoid defoliation created by Edipoda locusts and other living things, the Egyptians prayed to Sobek— the crocodile-headed god of animals and insects. As the son of Neith, Sobek was the underworld demon of the four elements—fire, earth, water, and air. At his cult center in Arsinoe (Crocodilopolis) where devotees cared for his sacred crocodiles, Sobek was ritually associated with Ra (fire), Geb (earth), Osiris (water), and Shu (air). It was believed that Sobek controlled such elements to the extent that he restrained the activity of certain creatures within specific habitats. Thus Sobek limited the activity of a crocodile within water, a locust within air, etc. His mastery of such elements was demonstrated in the Isis/Osiris myth when Isis searched the Nile for the severed body-parts of her husband/brother Osiris. Sobek, out of respect for the goddess, limited the appetite of the river animals and thus spared the floating pieces of Osiris.

As an Egyptian demon, Sobek was associated with the goddess Ammit—the crocodile-headed "eater of souls" that dwelt beneath the Scales of Justice in the judgment hall of Osiris. At other times Sobek and Ammit were depicted as one and the same. In his book, *Egypt, Gift of the Nile*, Walter A. Fairservis, Jr. paraphrased a section of the Book of the Dead. In the following paragraph he describes Sobek in the role of Ammit:

According to the book Meri would at last reach the place of the greatest test of all—the Great Judgment Hall. Here in the presence of Osiris, King of the Dead, Anubis the embalmer, Thoth the ibis-headed scribe, and the 42 gods of judgment, the heart of Meri would be placed on the scales to be weighed with the "Feather of Truth." If Meri had been a bad man in life, no words or deeds could save him now. His heart would not balance the Feather of Truth, and Meri would be thrown to Sebek (Sobek), the crocodile-headed eater of souls.

In another Sobek-related chapter of the Book of the Dead—Making The Transformation Into The Crocodile-God—we read:

The Osiris Ani, whose word is truth, saith: I am the crocodile god [Sobek] who dwelleth amid his terrors. I am the Crocodile-god and I seize [my prey] like a ravening beast. I am the great Fish which is in Kamui. I am the lord to whom bowings and prostrations are made in Sekhem.

Such bowings and prostrations were made to Sobek along the Nile river and at his centers at Kom Ombo and Thebes. Sobek's oblations often included human sacrifices and such may have been the Pharaoh's intention when he commanded the midwives to throw the Hebrew children into the Nile. The offerings anticipated Sobek's favor in delivering from bothersome insects, and, if a person wanted to eradicate an annoyance—such as locusts—they simply made supplications to Sobek while chanting, "To Sobek with it (the locust)!" The modern-day slang, "to Hell with it!" is a derivative of such ritual.

It was undoubtedly against the demon-god Sobek—and his pestilence-protection rituals—that the Hebrew God initiated the relentless plague of the locusts. In so doing, Yahweh revealed that Sobek was unable to control the elements, or limit the activity of God's insect army. Sobek's companion—the high god Ra (of fire)—could not scorch the creatures. Ra's son Shu—the Egyptian god of sun and wind (air)—could not blow the consuming insects away. It was not until the Hebrew God commanded "a mighty strong west wind, which took the locusts, and cast them into the Red sea" (Ex. 10:19), that the grievous plague was ended. Even so, the heart of Pharaoh was hardened against the God of Israel.

The darkness plague

"...and there was thick darkness in all the land of Egypt three days..."

My research companion, Dr. Jones, (I call him "Indy" after Indiana Jones) spoke recently of his trek through Hezekiah's Tunnel in Jerusalem. He described the interior of the cave as dominated by a darkness "...that could be felt, compounded by feelings of claustrophobia, obscurity, and utter dejection." One can imagine the terrors that the Egyptians must have experienced when the Hebrew God devised a proportional darkness that spread throughout "all the land of Egypt" and lasted for three days! Such an occurrence must have caused an unparalleled despondency, and most certainly would have devastated the Egyptian's religious idea that Amun-Ra ("The Hidden One") was the incarnation of the midday sun, and the most powerful god in the Cosmos.

The Egyptians referred to Amun-Ra as "the king of the gods." They believed that no deity was superior to him, and that the whole of the pantheon would perish without his symmetry. The sun itself was considered "the Eye of Amun-Ra", and the light and warmth of the midday sun was perceived as the bath of his blessing. Amun-Ra was also called Khepri (the rising sun), and Atum (the setting sun), so that each position of the sun—rising, midday, and setting, was perceived as a posture of Amun-Ra. According to myth, Amun-Ra, like the Sumerian god "Utu" (Shamash), traversed the sky each day. At night he journeyed through the underworld where the evil god Apepi attempted to prevent him from rising again. With the assistance of the magical masturbation-rituals conducted by the Egyptian priests, Amun-Ra was empowered each night to conquer Apepi and become the Ra-Harachte—the bright and morning sun. His cult center at Thebes was the primary location of such rituals, and the same site boasted the largest religious structure ever built—the temple of Amun-Ra at Karnak. Interestingly, the great Temple of Amun-Ra (with its 100 miles of walls and gardens) was the primary object of fascination and worship by the nemesis of Moses—the Pharaoh of the Exodus, Ramses II. It was believed that each pharaoh, including Ramses II (who completed Amun-Ra's temple), reconciled his divinity in the company of Amun-Ra during the festival of Opet. The festival was held at the Temple of Luxor and included a procession of gods carried on barges up the Nile river from Karnak to the Temple. The royal family accompanied the gods on boats while the Egyptian laity walked along the shore, calling aloud and making requests of the gods. Once at Luxor, the Pharaoh and his entourage entered the holy of holies where the king joined his ka (the mysterious ritual is unknown) and transmogrified into a living deity. Outside, large groups of dancers and musicians waited anxiously. When the king emerged supposedly "transformed," the crowd erupted in gaiety. From that day forward Egypt was "guarded" by their king and the Pharaoh was considered the son of the sun god—the earthly representative of the creator deity, Amun-Ra.

Subsequently, it was believed that the midday sun arose above Egypt because the Pharaoh had been honored and inaugurated in the Temple of Amun-Ra. If the sun was ever darkened or eclipsed, it was an evil omen for the king. Egypt's priests carefully interpreted such "signs," and even offered life-saving maneuvers to the Pharaoh. But when three days of utter darkness paralyzed the Egyptians (Exodus 10:21–23); the number three being understood by the Hebrews and the Egyptians as representing divine providence, the king's magicians were uncharacteristically silent. Like the three hours of darkness that accompanied the death of Christ (Luke 23:44), the sovereignty of the Highest was believed to be at work. It would do no good to call upon the goddess Nut. She had been proven powerless before Yahweh. If the God of the Hebrews was at work, Nut could do nothing to elevate herself nor could she force the light of Amun-Ra to shine. The sky-cow-goddess Hathor had been equally humiliated by Israel's Lord during the deadly murrain, and the evil god Sobek had been found impotent at controlling the element of sky. The mystical spells of Isis were useless against Yahweh. The priestly magic, paralyzed. And now, Amun-Ra, the Creator "king of the gods" and champion of the Egyptian pantheon, was confirmed helpless before the God of Hebrew slaves. "And Pharaoh called unto Moses, and said... Get thee from me, take heed to thyself, see my face no more; for in that day thou seest my face thou shalt die. And Moses said, Thou has spoken well, I will see thy face again no more" (Ex. 10:24; 28–29). With this final act of hardness, Pharaoh sealed the destiny of his kingdom, and, sadly, that of his firstborn son.

Death of the firstborn

"...And all the firstborn in the land of Egypt shall die, from the firstborn of Pharaoh that sitteth upon his throne, even unto the firstborn of the maidservant that is behind the mill..."

At least six deities were committed to the protection of Egypt's children. They included Heka, the mystical frog-goddess, who oversaw the development of animals and children beginning at the embryonic stage; Isis, the advocate-mother of the children who kept her word; Min, the god of virility who conferred reproductive vigor upon men and who was ritually called upon to produce an heir to the pharaoh; Horus, the son of Isis and Osiris, who protected the Pharaoh's son; Bes, the patron protector of mothers and their children; and the Pharaoh himself—Egypt's protector-incarnation of Amun-Ra and Horus. The female deities, Heka and Isis, oversaw different aspects of the children's physical development, while

Min and Horus were the powerful male deities responsible for the spiritual progress and overall health of the child.

Min's full name was Menu-ka-mut-f ("Min, Bull of his Mother"), and he was often worshiped in the image of a white bull. At other times Min was depicted as a bearded man with an oversized phallus. Such iconography of Min served to verify his position as the eminent Egyptian god of male sexuality, while also accounting for his mythological marriage to Qetesh—the equivalent Egyptian deity of female sexuality. Egyptian boys supposedly acquired their sexual strength from Min, and subsequently made offerings of lettuce (considered an aphrodisiac by the Egyptians) to this god. The Greeks confused Min with Pan—the Dionystic god of unbridled sexual desire—and thus participated in the orgiastic festivals held in his honor. But the most important area of Min's dominion, insofar as the Pharaoh was concerned, was the mystical relationship between the god and the royal family, including Min's association with the princely heir of Egypt—the pharaoh's son. The pharaoh was so concerned with the blessings of Min that he ceremoniously hoed the lettuce fields during the festival of this god. The idea was to humble himself in the presence of Min and thereby procure divine favor and reproductive synergy. Sexual energy, such as was abundantly produced by Min, was believed to be synonymous with health and longevity. Thus, if the pharaoh and his son were to live long and prosperous lives, they required the favor of Min—the preeminent god of sexual power. Such power of Min would have likely been sought during the death of Egypt's firstborn.

Legend has it that the god Horus was also involved in guarding the pharaoh's son, due, in part, to the mythology that the child Horus had been subjected to homosexual rape by the evil god Seth. The adult Horus was thus protective of children in general. Equally important, Horus was believed to incarnate himself within the living pharaoh, and to fill the heart of the pharaoh with respect for the father. The virtue of such parental respect was an important part of ancestor ritual, and referred to the story of Horus and his war with evil Seth over the murder of his father. Such myth supposedly contributed to the survival of the pharaoh and his son in two important ways: 1) Horus was the protector of the father and child, and perched above and behind the pharaoh, spreading his wings around and guarding the pharaoh's head (another plagiarism reminiscent of the Old Testament passage "in the shadow of Thy wings"); and, 2) Horus reminded the royal son of his responsibilities toward the father, especially of the offerings to be made daily at the deceased father's tomb. Such offerings were deemed necessary for the maintenance of the afterlife, and amulets (the eye of Horus) placed beside the offerings protected the stomach of the dead. In this way the living pharaoh (Horus) served the needs of the deceased father, while the predecessor pharaoh conducted himself as the Osiris in the underworld.

In the classic film by Cecil B. DeMille, *The Ten Commandments*, Yul Brynner, in the role of the pharaoh, placed his firstborn son in the arms of the falcon-headed god, Seker (who protected the dead as they passed through the underworld), and said, "Seker, great lord of the lower world, I...bow before you now. Show that you have power above the God of Moses. Restore the life he has taken from my son. Guide back his soul across the lake of death to the place of living men." Ramses II undoubtedly prayed in such fashion for the life of his son. Nevertheless, "at midnight the Lord smote all the firstborn in the land of Egypt, from the firstborn of Pharaoh that sat on his throne unto the firstborn of the captive that was in the dungeon" (Ex. 12:29). By initiating the death of the firstborn, Yahweh executed His final judgement "against all the gods of Egypt" (Ex. 12:12). Heka was proven powerless. Isis was defunct. Min was unable to energize the pharaoh's son. Horus was equally inept. The pharaoh was without a successor to watch over his tomb. Amun-Ra was without earthly representation. Egypt was without an heir. And the whole of the Egyptian pantheon, with its magic, myths, and rituals, crumbled at once beneath the feet of the Hebrew God. "And Pharaoh rose up in the night, he, and all his servants...for there was not a house where there was not one dead. And he called for Moses and Aaron by night, and said, Rise up, and get you forth from among my people, both ye and the children of Israel; and go, serve the Lord, as ye have said" (Ex. 12:30–31).

As noted in my original *Gods Who Walk Among Us*, from which the list above is derived, many read the Exodus account without even considering that the plagues might have been a domino effect from first to last, each one inseparably linked in a sequential order as judgments against specific Egyptian gods or goddesses (and a way for Yahweh to show His power was greater than theirs), and also miss the question of why the plagues were carried out in this specific succession, or why there would only be ten judgments when the exceedingly polytheistic Egyptian belief system involved far more than only ten deities. It's common to view God's wrath as more of a random series, basing the arrangement of disasters on His prerogative alone and not necessarily on a scientific and natural sequence of events that all stem from the initial judgment: a river of blood. Yet though God *frequently* defies the laws of science and nature throughout the Bible (and still today), proving that *He is never bound by such laws*, there are times when He will use the natural, scientific progression of events to accomplish His will, since He is the Creator of even science.

Exodus may well have been one of those occasions...and if it *was*, then it changes everything we *think* we know about Wormwood's looming catastrophe.

A documentary called *The Exodus Decoded*[13] shows the possible chain reaction. Before we get into that particular trail, though, one blood-versus-gas elephant in this room needs to be addressed: The crew behind this particular source suggests that the water-into-blood plague may have been CO_2 gas secretions making the waters appear to be as red as blood, and therefore their approach to the first plague is more symbolic than all the others, which they variably treat as literal. This inconsistency aside, however, their argument is supported by two gas-leak natural disasters—Lake Monoun in 1984 and Lake Nyos in 1986, both in Africa's Cameroon—from which the plagues of Egypt were repeated, one by one, astonishingly close to the biblical account.

In case what I just wrote wasn't completely clear, I will reiterate it: According to the documentary film *The Exodus Decoded* (aired August 20, 2006, on the History channel from Israeli-Canadian filmmaker Simcha Jacobovici and producer-director James Cameron of *The Terminator, Rambo, Aliens,* and other box office mega films fame) the ten plagues of Egypt—just as in the book of Exodus *and* in the same exact order—occurred twice in small villages in the 1980s. We have photos, testimonies, and reliable historical documentation of these disasters. (Note that because the Lake Nyos tragedy was more substantial than the other, most of our focus will be related to that event.)

Both tragedies began when iron-rich water from the deepest crevices of the lake floor was released via earthquakes (a "limnic eruption" or "lake overturning"), mixed with the rest of the water, formed an iron hydroxide (basically rust), and turned the color of the entire lake from blue to a deep red.

Have you ever bitten your tongue and tasted that iron-metallic taste? Have you ever smelled blood and equated it with that wet-metal smell? If so, you can probably see how the Israelites/Egyptians may have described a high-iron, gas-poisoned, blood-colored water substance as blood, *if* that was what they were dealing with. It wouldn't be much different than other areas of the Bible where something scientific (such as Earth orbiting around the sun) was described in words that the writers found familiar (such as the rising and setting of the sun in Psalm 113:3).

Without a doubt, many followers of the Abrahamic faiths will believe the Exodus narrative to be describing a literal blood, as opposed to any gas-leak phenomenon that may have occurred as a result of the volcano-induced earthquakes happening in the same place at around the same time.[14]

Our purpose in this study is not to argue this blood-versus-gas issue one way or the other. Nor is it to accept the documentary's explanation on all, or even any, of the plagues of Egypt.

We, with an acute focus on Wormwood, are interested only in discussing a hypothetical scientific and natural approach to what could happen after the waters are essentially rendered poisonous and unlivable/undrinkable, which in Egypt could have happened *regardless* of whether it was blood or gas.

Just as a quick reminder, here is the order of the plagues of Egypt:

1. Water into blood

2. Frogs

3. Gnats (or lice)

4. Flies

5. Death of livestock

6. Boils

7. Hail

8. Locusts

9. Darkness

10. Death of firstborn

What might have occurred in Egypt from a natural-disaster purview? Using the Cameroon tragedies, we can retrace and then project the following:

Once the oxygen in the lake water is stripped to below the point of survival for underwater animals as a result of the first plague, they die, and their corpses begin to rot. If the water wasn't already poisonous by this time, it certainly would be now. Unlike fish, frogs have the ability to leave the water and hop around on dry land, fleeing from the suffocation of aquatic pollution, explaining the second plague. Winged pests attracted to the smells of death and rot would logically swarm any people and/or animals living in close proximity to a lake filled with dying (or already deceased) fish. Without clean water, bacterial outbreaks drawing gnats, lice, and flies occur, explaining the third and fourth plagues. The live-stock, vulnerable to both the contaminated water and the disease carried by flies, and more, begin to drop one by one in a death epidemic matching proportions of plague five. The enormous skin blisters found on the human and animal bodies (both corpses and survivors) of the Lake Nyos residents in 1986 shows that continual and intense exposure to decomposing flesh, poisonous waters, and disease (as well as acidic gases in the air) can most certainly lead to what we see described as boils in the sixth plague.

From this moment, makers of the documentary rely on the idea that the volcano that erupted around the time of the Exodus story would have released ash high into the air nearby. Volcanologists acknowledge that when cooler sections of this ash cloud accumulate with atmospheric moisture and water vapors, it creates a type of hailstone called a nucleus. This, in addition to the fiery fallout of the eruption, creates *accretionary lapilli*, or "volcanic hail,"[15] giving enlightenment to the otherwise impossible weather phenomenon of plague seven. If this sudden shift in climate conditions was in fact occurring at the time and in the area of the plagues of Egypt, then it stands to reason that locusts—which corporately relocate en masse in heavy, flurrying numbers up to "between forty and eighty million adult locusts in each square kilometer" toward heat[16]—would ravage the Egyptians and account for plague eight. Plague nine, darkness, is explained by *The Exodus Decoded* as the ash clouds of the eruption finally reaching the air around the Nile River and casting the land in thick shadows.[17]

The documentary discusses what occurred next in 1986 at the edge of Lake Nyos: the gas leak at the bottom of the waters that started all of these events (approximately six months prior) finally reached the surface, releasing a carbon dioxide fog that traveled across the top of the water, sweeping its deadly mist onto the nearby land. For the better part of a night the toxic fog remained in the air nearest to the ground, as the fog is heavier than air. The gases eventually rose up and dissipated, becoming harmless, but *not* before they claimed the life of any person who went to bed that night with a close-to-the-ground (or floor) bedding arrangement. The following day those who slept from elevated positions woke to find almost eighteen hundred people dead.[18]

According to *The Exodus Decoded*, this is the same anomaly that brought the tenth plague of Egypt to fruition: "Egyptian first-born males had a privileged position. They were the heirs to the throne, to property, titles, and more. They slept on Egyptian beds low to the ground while their brothers and sisters slept on rooftops, sheds, and in wagons. The Israelites, sitting up at their first Passover meal, did not feel a thing, while the low-traveling gas suffocated the privileged, Egyptian males, sleeping in their beds."[19]

What a fascinating theory. For certain, it's the only plausible scientific or natural explanation that this author has easily found for the plague of the firstborn deaths.

Eruptive phenomena

We started this Egypt-connection section of this study stating that God is not bound by any scientific or natural law, that He *frequently* overpowers them, and that *if* those laws happen to apply in a biblical narrative of divine intervention, it's because God chose to use the universe and laws He created to execute His will. Many people become angered at the suggestion that natural elements can explain God's wrath, plagues, or miracles, because they feel as though this limits the Lord. However, God is no more limited by creation than a master chef is restricted by the professional cookware he uses. These are elements below Him, created *by* Him (although for argument's sake I will concede that the chef does not usually invent/create his tools) for His own pleasure (Rev. 4:11), and it is His prerogative to use them in any way He sees fit. Consider the Book of Jonah and God's mastery over His own creation, even as the man He called defied the will (and orders) of the almighty God. Throughout the entire story we are shown that God utilized various elements of His creation in order to exact His will:

- "But the LORD sent out a great wind into the sea" (Jon. 1:4).

- "Now the LORD had prepared a great fish to swallow up Jonah" (Jon. 1:17).

- "And the LORD spake unto the fish, and it vomited out Jonah upon the dry land" (Jon. 2:10).

- "And the LORD God prepared a gourd" (Jon. 4:6).

- "But God prepared a worm...and it smote the gourd" (Jon. 4:7).

- "And it came to pass, when the sun did arise, that God prepared a vehement east wind" (Jon. 4:8).

This story shows that even in the simplest elements of creation (such as a gourd or a worm), the Lord carefully orchestrates the interaction of His creation. Likewise the plagues of Egypt hold too many potential variables to go down the way they did without His orchestration. If even *one thing* went wrong—such as a little Egyptian girl deciding to sleep on the floor the same night of a gas fog and dying before morning even though she wasn't a first-born son—Pharaoh's people could have understandably said that God made a mistake in carrying out His judgments. I believe the Word when it says that the angel of death passed through Egypt that night, even if that messenger did in fact coordinate the spread of fatally toxic gas.

According to *The Exodus Decoded*, this chain of "plagues" on a few small villages in Cameroon in the 1980s started because of the volcanic eruption nearby. As stated, this accounts for both the deep-water limnic eruption (or lake overturning) that released the gases into the lake and the hail-and-fire phenomenon. If the time-line of the Exodus account follows that which the documentary makers suggest, then the Santorini volcano could have easily been the genesis calamity that kicked off all the subsequent plagues of Egypt. (Interestingly experts behind this film found ash grains of the Santorini volcano in the Nile River dating to the time of the Exodus, which supports the water-into-gas-leak-blood theory, as well as the idea that darkness enveloped the land through thick, penetrating ash clouds.)

Follow this thread of thought for a moment...If this volcano theory holds clout for a future Wormwood scenario, then the actual poisoning from the bitter waters *might* be brought on by seismic activity that releases toxic gases into our freshwater sources.

"But Tom," readers might be thinking, "are you suggesting that Wormwood is actually a volcano instead of the falling star that Revelation describes?"

Not exactly. There is one fact about asteroids or falling stars that now demands to be brought to the forefront of this discussion: scientists openly acknowledge that asteroid or comet impact, if the space rock is large enough, can certainly generate volcanic eruptions. The abstract from the *Science Advances* article "Anomalous K-Pg–aged Seafloor Attributed to Impact-Induced Mid-ocean Ridge Magmatism" stated in February of 2018: "Eruptive phenomena at all scales, from hydrothermal geysers to flood basalts, can potentially be initiated or modulated by external mechanical perturbations....Magmatism on a global scale [was likely triggered] by the Chicxulub meteorite...[as] recorded by transiently increased crustal production at mid-ocean ridges."[20] To clarify: "External mechanical perturbations," in the context of this particular article, means asteroids, comets, and so on; "global scale" means that the volcanic activity happened all over the planet at once; and the "Chicxulub meteorite" is the meteorite scientists say is responsible for the extinction of dinosaurs. In other words, scientists have identified a link showing that one enormous space rock of the past caused volcanoes all over the world to erupt, leading to the extinction of one of the strongest species of animal God ever created. Earth is also not the only planet that has been coupled with this impact/volcano association. As *Scientific American* acknowledges, other planets within our solar system display "giant craters caused by impacts [where] there's evidence of volcanic activity."[21] (On a separate note, Donald Patten, author of the previously discussed work titled *The Biblical Flood and the Ice Epoch*, likely referenced the Chicxulub impact, although this was never given a specific name in his work. It is thought that he avoided naming this particular event as the culprit of the biblical flood due to the fact that some scientists date the Chicxulub catastrophe as far back as 66 million years,[22] while Patten's own dating of the event was much more recent. This discrepancy opens the ever-ongoing debate of evolution vs. creationism which is beyond the scope of both his book and this one.)

Interestingly Revelation's *second* trumpet—described as the "great mountain burning with fire" that turns the water to blood and kills one-third of all sea life and ships (8:8–9)—is frequently thought to be a volcano. First, the description matches identically since a volcano is literally a mountain exploding with fire. Second, volcanic eruption can and has resulted in colossal sections of a mountain's mass separating and "casting itself" off of its main body, so even though the rendering of verse 8:8 sounds like something that is falling from the sky, it wouldn't necessarily have to be airborne for a "great mountain burning with fire" to be "cast into the sea." This is, again, just another guess, but it's astonishing how so much of these trumpet studies tend to come back to the idea that we might be dealing with some kind of eruptive phenomenon, that a star wouldn't have to be carrying any kind of poison on or within itself to essentially bring ruin to one-third of *either* salt or fresh waters.

So no, my theory is *not* that Wormwood is a volcano. My theory here is that the impact of the falling star Wormwood could set off another group of limnic eruptions that release the water-into-blood CO_2 catastrophe and perhaps even awaken surrounding volcanoes that exacerbate the entire cataclysm of events.

In any case…might we be facing a time when the third trumpet judgment, Wormwood, brings back the plagues of Egypt? Those dreaded, awful plagues that have haunted the imaginations of believers for thousands of years—images of those bloody waters, pests crawling in and out of every bodily orifice, painful sores that ooze fluids, the odors of death and decay…Will Wormwood be the catalyst of a repeat, Exodus-like judgment, but this time on a *global scale*?

Although not every scholar agrees with the conclusions of the documentary, a connection is still irreversibly made here, and by two historical "ten plagues like Egypt" events (Lake Nyos/Monoun) that nobody can deny actually happened. Whether or not the divine intervention or display of God's power was enacted *in Egypt* the way *The Exodus Decoded* shows, that *was* the way the natural order of things went down in Cameroon.

Suddenly, from this angle the ten plagues of Egypt become extremely relevant to the discussion of the seven trumpets, since they parallel very similar destruction/judgment descriptions. And suddenly, from this angle the "poisoned waters" of Wormwood take on a new face of death. Once this Egypt-Wormwood link is made, one can't view the bloated, infected bodies scattered about the land near the lakes of Cameroon without feeling a connection to the mortality of tomorrow. The testimonies, like that of Lake Nyos survivor Joseph Nkwain, give us a more personal, intimate, frightening, and *real* glimpse of what countless people may face after Wormwood strikes and their poisoned bodies begin to fail:

I could not speak....I could not open my mouth because then I smelled something terrible....I heard my daughter snoring in a terrible way, very abnormal....When crossing to my daughter's bed...I collapsed and fell....I was surprised to see that my trousers were red, had some stains like honey. I saw some...starchy mess on my body. My arms had some wounds....I didn't really know how I got these wounds....I opened the door....I wanted to speak, my breath would not come out....My daughter was already dead....I went into my daughter's bed, thinking that she was still sleeping. I slept till it was 4:30 p.m. in the afternoon...on Friday. (Then) I managed to go over to my neighbours' houses. They were all dead....I decided to leave...(because) most of my family was in Wum [a nearby village]....I got my motorcycle....A friend whose father had died left with me (for) Wum....As I rode...through Nyos I didn't see any sign of any living thing.…(When I got to Wum), I was unable to walk, even to talk…my body was completely weak.[23]

In a BBC News article published only days following the natural catastrophe, victims were subject to some extremely graphic horrors, as one doctor at the hospital in Yaoundé reported symptoms including "burning pains in the eyes and nose, coughing, and signs of asphyxiation similar to strangulation, as like being gassed by a kitchen stove."[24] These respiratory tortures were, of course, in addition to the skin blisters or *bullae* (Latin for "bubble"; in other words, "boils") that wreaked havoc on the body from head to toe on many of the survivors. A follow-up report released in the *British Medical Journal* on May 27, 1989, by Drs. Peter J. Baxter, M. Kapila, and D. Mfonfu states that even among those five thousand in the area who fled to safety early on, many still suffered intense lesions, permanent skin scarring and discoloration, permanent respiratory conditions, and even paralysis.[25]

At the very least, whether Wormwood's trail of desolation treks the way Egypt or Lake Nyos (or any other place) ever has, we have been given a sneak peek into a future reality that shows an escalation of tragedy *far* more serious than some have assumed. When Wormwood hits, the challenge won't just be to find clean water to drink. That poison is going to trickle into every crack and crevice of surrounding society, permeating everything it touches with infection, disease, torment, and death, maybe even for those who never ingest a drop of the water Wormwood pollutes.

Boiled arsenic?

Don't assume that the CO_2 theory is the only method by which a falling star could poison water. It's merely one way that took a bit longer to explain. As recently as September 15, 2007, we have observed what kind of poison gases can bleed up and out of otherwise ignored soil in the middle of almost nowhere. All it took was a seven-to-twelve-ton chunk of chondrite space rock traveling at 27,000 miles per hour and heated to three thousand degrees Fahrenheit, and local villagers were struck with extreme illness, the symptoms of which were so wide and seemingly disconnected that it initially looked hopeless that they would isolate a cause or cure. This rock was only the size of "a dinette set," so its water-poisoning potential on the "empty plain" in Carancas, Peru, is obviously incomparable to any sizable trumpet judgment that enters large bodies of water or moving streams. Nevertheless, *within only hours of collision*, mysterious sickness had spread to the point that the locals began to whisper that the meteorite's scattered debris was, among other theories, "cursed."[26]

Cursed—the true, etymological, Hebraic meaning behind the name of the Revelation star that will cause many people in the end days to fall ill and die. The word sounds illogical, sensational, and maybe even childish, but when the clinics around Carancas were filling with vomiting patients (including policemen) and news reporters were scrambling to get the story on the bizarre nosebleed outbreak of nearby cows and livestock—again, only hours after impact—suddenly *cursed* seemed like a valid adjective.

Before long, symptoms increased in intensity, and the numbers of locals who were becoming sick grew. Medical tents had to be set up intermittently throughout the area as the original one hundred or so patients increased to six hundred; meanwhile, livestock started to die, according to one report.[27]

Then, luckily, *wonderfully*, just as it appeared that it couldn't get any worse, it suddenly got better—and fast. An explanation for the illness was offered: "Arsenic in the water table had been heated and vaporized by the impact energy and sent into the air as a gas."[28] Putting it another way, "engineer Renan Ramirez of the Peruvian Nuclear Energy Institute [stated]...'It is a conventional meteorite that, when it struck, produced gases by fusing with elements of the terrain.'"[29]

Arsenic poisoning...who would have thought?

Within days, after several experts had a chance to weigh in on the subject, the arsenic angle was confirmed: a blazing hot meteorite, itself high in iron and magnetic energies that would assist in maintaining its temperature, embedded itself in a dry plain, immediately drawing up water from the earth, boiling toxic elements already present in the water/soil, and releasing the harmful concoction into the air, into which it eventually dissipated.

As it turned out, no ground, rocks, or space debris were cursed after all, and the spike in noxious sickness was never related to anything anyone touched or drank. Thankfully no humans lost their lives, and soon after the ordeal the impact of Carancas' chondritic meteorite, for the most part, slipped into obscurity.

What it *will* always be remembered for, however, is a highly unusual poisoning event that nobody could have seen coming or planned for. Nobody would have guessed that there was arsenic present in that particular soil, that this exact location would be hit with a meteorite that also happened to be able to preserve its own temperature in Earth's atmosphere, and that the boiling of moisture beneath the surface of the dirt would melt and distribute toxins. Had that been an asteroid the size that Wormwood will likely be— and had it driven into the soil layers of a major lake or river where, currently unbeknownst to us, there are existent toxicities—then drinking that water could result in the death of what Revelation refers to as "many," certainly. But it also wouldn't require rattling any limnic eruptions loose to bring fatal gases into our atmosphere, potentially causing similar catastrophic results to the Egypt and Lake Nyos scenarios we have discussed up to this point.

We don't know exactly how this "star" of Revelation is going to poison our fresh water, but the ten worst-nightmare plagues of Egypt are certainly a possibility.

This time, however, God will be raining down judgment on more than just a pharaoh and his pagan gods.

THE IMPACT OF IMPACT

Even more alarming when we consider Wormwood is the death rates we may be facing in those coming days. The Lake Nyos tragedy claimed 1,746 human victims and an estimated thirty-five hundred animals in a sixteen-mile radius of the lake's shore. In a population of approximately six thousand in the affected villages, that calculates to a death rate of, ironically, about one-third. (There's that *one-third* fraction again.) As absolutely heartbreaking as these numbers are, they pale in comparison to the losses the Egyptians might have experienced in what most scholars agree was a population of three or four million, and they may not have been able to flee for refuge to an unaffected area as the Lake Nyos victims did. Additionally the edge of Lake Nyos affected sixteen miles, whereas the Nile River of Egypt would have potentially reached up to 750 miles, assuming that the poisoning influenced life across all the general boundaries of ancient Egypt (which stretched "from the Nile's mouth at the Mediterranean [the north end] to Aswan [the south end]"[30]). *Not* counting the deaths of every firstborn in the tenth plague, and *not* counting the obliteration of the Egyptian military in the crossing of the Red Sea, the casualty count of the pharaoh's people by Nile poisoning alone might have been eventually approximately 1,060,606 people. (This calculation is based on one-third of a 3.5 million population, the one-third pattern following Lake Nyos.) Obviously that is just a guess stemming from a theory that the Exodus account involved deaths that the Bible didn't detail. Not only is that extremely likely based on the natural result of poisoning an enormous population of people, but one verse even hints at the idea that the Egyptians had already faced vast death numbers and were afraid of *complete* annihilation (Exod. 12:33), and even the classic works of Philo acknowledge that there was "a great multitude of people killed."[31]

Let us take what we have studied so far and make a *purely hypothetical* guess at what we may be looking at here. If we were to suppose that the asteroid Apophis was *the* eschatological Wormwood star, then the third-trumpet judgment would, if planetary defense experts are correct, take place in the very near future—April 13, 2029. The world population is expected to reach 8.5 billion by that year,[32] *but we cannot assume that number would still be accurate* if all the seven seals had occurred already as well as trumpets one and two. If the death tolls of those judgments are collectively applied to the proposed 8.5 billion, we would obviously arrive at a smaller number. How much smaller at this time is impossible to know, but assuming the judgments come close together, it still gives us a starting point from which we can toss around some numbers and produce our earliest guess regarding the lives that might be affected by Wormwood's impact.

As mentioned briefly earlier on, half of the earth's human population will have been eliminated between the fourth seal and the sixth trumpet.[33] If we start with an estimate of 8.5 billion at the fourth seal and end with 4.25 billion at the sixth trumpet, with nine eschatological judgments/events between them, it equals an average of 0.47 billion deaths per event (half the population divided by nine events). There are six events between the fourth seal and the star Wormwood, which brings our hypothetical population down to 5.7 billion by the third trumpet (0.47 billion times the six events [2.8 billion], extracted from the starting point of 8.5 billion). Hold on to that number for a moment.

It's also impossible to discern precisely *where* on the earth Wormwood will hit in order to contaminate one-third of the planet's freshwater supply, and its point of impact will not be anywhere on our current world maps because of the sixth seal judgment carried out prior: "And I beheld when he had opened the sixth seal, and, lo, there was a great *earthquake*; and the sun became black as sackcloth of hair, and the moon became as blood; and the *stars of heaven fell unto the earth*, even as a fig tree casteth her untimely figs, when she is shaken of a mighty wind [consider this to be a kind of insane meteor shower]. And the heaven departed as a scroll when it is rolled together; and *every mountain and island were moved out of their places*" (Rev. 6:12–14, emphasis added). Our current maps will be irrelevant after "every mountain and island" has shifted, earthquakes have redirected so many of the rivers, meteors have disturbed the glaciers, and so on. We can't even venture a guess at this point as to where our freshwater sources will be, where the fresh water would flow, and where that might put the one-third that Wormwood will infect, or even how much clean water still exists and exactly how much that one-third would represent in liquid mass. (Interestingly all this activity will most certainly result in eruptive phenomena.)

However, following the patterns we've used up to this point, relying on what we've ascertained from the Lake Nyos event (remember, one-third of the people died), and assuming that the majority of the world's remaining population lives near the one-third of the polluted fresh water, we arrive at one-third of 5.7 billion eventually meeting their Maker by the influence of Wormwood-poisoning *alone*.

This judgment will be, as the root word of *wormwood* means, truly a curse.

PICTURE THIS...

When the first trumpet (hail and fire mixed with blood) rains down upon the earth, there will either be a quick scientific explanation, if

God chooses to use His natural elements, or there will be a lot of scrambling to figure out what unexplainable weather phenomenon could have caused something so unusual, if God chooses to allow literal blood mingled with ice and fire to fall from the sky. If the latter of these two scenarios is correct, then I predict an "answer" will be produced, announced, and broadcast all over the world as soon as possible just to keep the masses calm and curb potential chaos and rioting. The "answer" that the experts (probably the Antichrist's crew) give, in that case, will most certainly be falsified. Authorities will craft what sounds like a completely reasonable rationalization for how the incident occurred and make sure their story is carried to the ends of the earth and acknowledged as the truth. By the time a few brilliant and educated minds stand up to challenge the holes in the world government's explanation, too many people will have accepted the lie, and the minority will be written off as crazed conspiratorialists. Meanwhile those who have been waiting for the signs of the coming of the Son of Man will see the weather phenomenon (depending on their dispensational leanings) for what it is and understand that they are looking at a mere continuation of the seals on into the first of the trumpet judgments, and the secularized world will be swallowing every scientist's cover-up and scoffing at "religious lunatics."

Trumpet two could go a number of ways. It could be a massive volcano that throws *itself* (enormous hunks of its own mass) into the sea like one of the theories we talked about pages back, it could be an asteroid or comet, or it could be that God throws a literal mountain on fire from the heavens into the sea. Depending on how natural the symptoms of the occurrence appear, we could be looking at yet another season where the events of Revelation are explained away because "experts say what happened was…" But beyond doubt this trumpet's blast will echo with a rise in the "railings" of "religious lunatics"—those men and women who see the act of God's wrath for what it is. The nature of many men in those days will be to deny God or His works until the very end, despite the great revival that spawns from the two witnesses and the prophesied judgments. Some will see blood fall from the sky, followed by whatever form the second trumpet takes on, and they will harbor great doubt in their hearts and minds regarding the authorities' explanation (*if* the authorities are still trying at this point; the last two seals and the first two trumpets might happen in rapid succession and look so supernatural that their efforts are completely shifted to broadcasting the Antichrist's wonders and peace talks)...but they will spend more mental and emotional energy proclaiming *any* theory over the truth written in Revelation so they can avoid being accountable to the God of the Bible.

The trumpets begin with things falling from the sky, and it only escalates. Each one is exceedingly more frightening and threatening to our mortality than the previous, and it's *all* cosmic. It's all about the heavens raining judgment, and the damage done to our earth will be so severe by this point in the progression of Revelation that there will be no time to recover. Relief and rescue organizations, scientists, astronomers, astrologists, NASA, all governments of the world...*everyone* will be overwhelmed with the constant disaster reparation agenda that will not allow progress to be made before the next judgment renders it irrelevant.

This is not really terrible storms that we're talking about. It goes beyond figuring out who will be able to reach the victims of a hurricane such as Katrina. It's the beginning of the end of the world, the completely irreversible differentiation between the world we know now and the dystopian world of post-apocalyptia. Whether people follow the true God of Abraham, Isaac, and Jacob or not, it is my opinion that by the third trumpet few will still be doubting His existence. Many will be deceived into believing the lies and deceit of the Antichrist, but I trust atheism will likely be obsolete.

Then, enter Wormwood into the scene...

If there was *any* doubt left in the world that the seals and the trumpets had been mere weather phenomena—if there were any remaining hope up to this point that our beautiful green and blue planet could eventually continue with life as we knew it—the poisoning of one-third of the earth's waters is the last sign before things get demonic and supernatural *real* fast.

We may be living as it was in the days of Pharaoh. And as frightening as that thought would have been at any previous moment in history, by the time it plays out the way it might in the future, the "curses" upon Pharaoh will look like small beans.

THE END OF DAYS

Scripture makes no small point about it: the third trumpet will be devastating and terrifying. It may be delivered via a meteor collision. It could come as a sweeping famine. A nuclear event is not entirely out of the question. Perhaps an asteroid will strike a nuclear power plant, and the ensuing desolation will launch a dearth unlike any the world has yet seen.

Or perhaps the event will have a different, darker undertone.

Maybe the third trumpet—or *all the trumpets*—will be a by-product of spiritual warfare taking place in heavenly realms, between which our planet—along with many other unsuspecting cosmic bodies—will become caught in the crossfire. Some may think that this possibility is surely an exaggeration, but consider this: we often make the assumption that since angels can appear in a fleshly, human form (Heb. 13:2), and since Jesus walked the earth as a man, in the flesh (1 Tim. 3:16), entities within the spiritual realm are of similar size and capability to human beings. But nothing could be further from the truth. We know from Scripture that God's strength is unending (Job 9:4), His knowledge is infinite (Ps. 147:5), and all victory is His (Col. 2:15). Furthermore, Revelation 10 gives us insight as to what one particular angel appearing during this time will look like:

> And I saw another mighty angel come down from heaven, clothed with a cloud: and a rainbow was upon his head, and his face was as it were the sun, and his feet as pillars of fire: and he had in his hand a little book open: and he set his right foot upon the sea, and his left foot on the earth, and cried with a loud voice, as when a lion roareth: and when he had cried, seven thunders uttered their voices.
>
> —REVELATION 10:1–3

This one holy entity is so large that the Bible tells us he *literally* stands over the land and sea. His voice—as a roar of a lion—is enough to call forth the response of seven thunders. Thus, this being has the strength to provoke natural elements upon the earth. Revelation is filled with accountings of holy angels interacting with natural elements in the earthly realm out of obedience to God, but what of the fallen ones? What about those entities that were once similar to the angels of heaven but rebelled alongside Lucifer so long ago? Will these agents of evil, during the last days of planet Earth as we know it, attempt to manipulate created fixtures as well? Perhaps another round of planetary ping-pong lies in the future of mankind's story.

We know in that day Satan will be rallying his forces to wreak havoc upon the earth because he is aware that his time is short:

> Therefore rejoice, ye heavens, and ye that dwell in them. Woe to the inhabiters of the earth and of the sea! For the devil is come down unto you, having great wrath, because he knoweth that he hath but a short time."
>
> —REVELATION 12:12

It stands to reason that during these last days the battle between good and evil, which has raged on since the beginning of the world, would culminate into a climax: *the battle of the ages.*

To some this may seem far-fetched, especially since some people interpret the trumpets to represent more subtle means of judgment, such as famine or pestilence. Jumping to the conclusion that these instruments of judgment could really mean that bodies within the universe are being knocked around like ping-pong balls seems to some individuals to be quite a jump.

That is, until the next few trumpets are explored:

> And the fourth angel sounded, and the third part of the sun was smitten, and the third part of the moon, and the third part of the stars; so as the third part of them was darkened, and the day shone not for a third part of it, and the night likewise.
>
> —REVELATION 8:12

This verse, if taken in literal context, clearly illustrates that cosmic disturbance will be completely unprecedented, unlike anything our solar system has experienced. Furthermore, this upheaval will permanently change things on Earth. There is no reparation that can be made for the type of damage that will occur for our planet and its surrounding orbital bodies during this catastrophic event. The ramifications of this trumpet and the following are so severe that afterward an angel will soar throughout heaven, shouting woes to them that must endure the things that are yet to take place (Rev. 8:13).

> And the fifth angel sounded, and I saw a star fall from heaven unto the earth: and to him was given the key of the bottomless pit. And he opened the bottomless pit; and there arose a smoke out of the pit, as the smoke of a great furnace; and the sun and the air were darkened by reason of the smoke of the

pit. And there came out of the smoke locusts upon the earth: and unto them was given power, as the scorpions of the earth have power. And it was commanded them that they should not hurt the grass of the earth, neither any green thing, neither any tree; but only those men which have not the seal of God in their foreheads.

—Revelation 9:1–4

There is no question that at this point heaven and hell are engaged in full-scale warfare. Entities from the bottomless pit are called forth and given permission to torment those who have not been sealed with the seal of God. These disturbingly creepy creatures are described as small horse-looking beings, likely equipped with an exoskeleton or other "armor," heads that look as if they are endowed with a gold crown, hair similar to a woman's, and teeth like a lion's. These entities sound like the chariots of horses on their way to battle, and their scorpion-like tails will sting people, tormenting them for five months. Despite the fact that this anguish will make men wish for death, we are told that such mercies will be out of the question for these unfortunate individuals. We are told that these are direct servants of Apollyon, the ruler of the bottomless pit (Rev. 9:5–11). By the time the sixth trumpet sounds, the four angels "bound in the great river Euphrates" (Rev. 9:14) are loosed and accompany an army of "two hundred thousand thousand," (Rev. 9:16)—in modern mathematical terms, two hundred million— spewing fire, jacinth, and brimstone upon the earth, killing one-third of the population:

And the four angels were loosed, which were prepared for an hour, and a day, and a month, and a year, for to slay the third part of men. And the number of the army of the horsemen were two hundred thousand thousand: and I heard the number of them. And thus I saw the horses in the vision, and them that sat on them, having breastplates of fire, and of jacinth, and brimstone: and the heads of the horses were as the heads of lions; and out of their mouths issued fire and smoke and brimstone. By these three was the third part of men killed, by the fire, and by the smoke, and by the brimstone, which issued out of their mouths.

—REVELATION 9:15–18

While it is possible to interpret this verse as reflecting famine, sickness, and plague, as some do, it is interesting to note the specific constitution of their attack: fire, jacinth, and brimstone. When considering astral activity, the connection to fire seems obvious. Brimstone is reasonable to include in the cosmic realm as well, when one realizes that the phrase is a reference to rocky objects of sulfurous content.[34] However, jacinth is likely the most surprising element in this context. Its original word, transliterated *hyakinthinos*, means "of a red colour bordering on black."[35] So through this passage we literally see evil entities being unleashed upon the earth in numbers as large as two hundred million at once, and these, assaulting mankind, are armed with a stony or rocky, reddish-black, fiery, sulfurous composition. Additionally their mounts embody the ferocity of lions, spewing fire, smoke, and brimstone upon the earth.

Could it be that this army will be unleashed upon our sky, a massive cosmic attack of evil entities disguised as asteroids, plunging toward the earth in impact numbers reaching two hundred million? When people consider the fiery, rocky constitution of these evil warriors, some may assert that this position is unlikely, but it cannot be ruled out with complete certainty.

By the time Wormwood hits, Earth will have been repeatedly pillaged by war, ravaged by disease, and beleaguered by famine. The blood of martyred believers will taint the weary ground. Devastating earthquakes will have rearranged the shape of the terrain in areas across the world, and the moon and stars will all be jostled from their places. One-third of Earth's green surfaces will have been burnt beyond fertility, and one-third of the life in or upon the sea will cease to exist. Previously constant, reliable elements such as day and night will now operate unhinged as days are darkened by the aftermath of cosmic upheaval unlike anything mankind has previously seen. And after Wormwood strikes, the supernatural attack upon humanity will become *more* blatant at the arrival of demonic locusts that are released from hell for the purpose of making mankind suffer, followed by two hundred million stony, fire-and-brimstone-equipped attackers who assault our planet—be it via asteroid, famine, warfare, pestilence, or other means—with one last attack before the mystery of God is completed (Rev. 10:7).

> The great day of the LORD is near, it is near, and hasteth greatly, even the voice of the day of the LORD: the mighty man shall cry there bitterly. That day is a day of wrath, a day of trouble and distress, a day of wasteness and desolation, a day of darkness and gloominess, a day of clouds and thick darkness, a day of the trumpet and alarm against the fenced cities, and against the high towers. And I will bring distress upon men, that they shall walk like blind men, because they have sinned against the LORD: and their blood shall be poured out as dust, and their flesh as the dung. Neither their silver nor their gold shall be able to deliver them in the day of the LORD's wrath; but the whole land shall be devoured by the fire of his jealousy: for he shall make even a speedy riddance of all them that dwell in the land.
>
> —ZEPHANIAH 1:14–18

When the trumpets begin to sound, a dark day has indeed arrived for planet Earth and all those who inhabit it. Death, disease, warfare, pain, torment, and dread will run rampant across all those who dwell therein on that terrifying day. The only hope for mankind in that moment will be found in whether they are one of

God's own. For these people a beautiful and joyous reunion with their Maker awaits them:

> And God shall wipe away all tears from their eyes; and there shall be no more death, neither sorrow, nor crying, neither shall there be any more pain: for the former things are passed away.
>
> —REVELATION 21:4

For those who reject Him, the nightmare will be just beginning...

Chapter 7

FICTIONAL CATACLYSM NARRATIVE

DANNY LEWIS SMACKED another twenty-dollar bill on the countertop between him and the elderly Hispanic restaurant owner. Jabbing his dark-skinned index finger down on top of the cash, he leaned in closer and spoke with a desperate, but whispery, tone. "Come on, man! This is all I have. Every last dollar. That's eighty bucks right there!"

"No, no, no. We no have food here." The man shook his head firmly, pointing to a crude handwritten sign in black marker perched in front of the cash register that said NO FOOD. He repeated himself in the same heavy accent, almost yelling: "We no have food here! You go other place!"

There was no "other place," and everyone knew it. It was always the same. Either the restaurants were completely empty, or the owners were the only ones left, hanging out at the front desk as if they intended to welcome paying customers while simultaneously turning all of them away. Occasionally a person could find a small café still in full operation, but judging by the rough clientele that frequented the joint, it couldn't be assumed just anyone was allowed to stroll in. Once when Danny so much as leaned in to peer through the glass, he had so many bikers stiffen and stare him down that he spent the next two hours hiding in the bushes for fear he would be stabbed if he tried to run. No, there was no "other place."

Danny's eyes shifted to the distracting television blaring in the back room. He recognized the familiar face of that weasel Steve Campion, who at this moment appeared to be flapping his jaw for the thirtieth time about how the #CitiesofRefuge leak was a hoax that nobody need pay any attention to. Behind him stood Dr. Gale Stone, obviously waiting her turn to give another round of promises that the government would leave no man behind in the case of a major city emergency evacuation. Behind *both* of them was a sunshiny White House, weather that DC hadn't seen in what seemed like forever, so Danny wondered how long ago this "live" statement was filmed for a later release to the rest of the country. To add to the dubiousness of the clip was the fact that the White House was visible at all. Everyone in this area knew that the White House had been completely barricaded for months.

Suddenly angered by the sight of Campion's face, Danny swatted at the NO FOOD sign, knocking it to the floor. The other man's fists tightened, whitening at the knuckles, so Danny switched to confident pleading. "Dude, I *know* you're hooking people up with grub! Don't play me like a fool. I got money! I'll pay you, man!"

As the man behind the counter flew into a Spanish tirade, Danny glanced over his shoulder to the only other person in the room: a disheveled and soaked older woman with wiry salt-and-pepper hair, dressed in blue teddy bear pajamas and bright yellow galoshes. She was standing in a hunched position, attempting to cram what looked like a small wad of tortillas into a purse overflowing with other food-filled baggies. As her head was down and she struggled with the last inch of a clearly broken zipper, she lurched forward toward the exit, accidentally ramming her head hard into Danny's rib.

"Watch it!" he screamed at her. She stumbled backward in fear, and they locked eyes.

"Sorry, sir," she said, holding a hand up defensively. "Please, no fight." Her accent was thicker than the other man's, and though it was hard to tell from only a few words, the way her voice hung on her consonants, she sounded distinctly Russian. "Please…"

Danny softened when he saw the layers of terror and panic emanating from her watery, bloodshot baby blues. Moments earlier she appeared to him like a filthy, frenzied, humpbacked hoarder creeping around the streets of Washington, DC, just to rob everyone else of their chances of getting a meal that day. But face to face, her chin quivering helplessly, there was something far more human about her—something perhaps even maternal. He looked down at her purse, and she clutched it protectively to her chest in response. For all he knew, whatever was in that bundle had been bartered to ensure the feeding of twenty-five starving children later that night.

Maybe this woman was one of those Rhode Island Avenue shelter mothers he'd heard about stemming from the relief arm of the resistance called the Patriots. Locals said these women would go to far and great lengths—some of which Danny couldn't even stomach hearing about—to feed the other Patriots' kids who'd been separated from them as a result of their activism in various protests, rallies, marches, campaigns, and so on, against Washington. These women wouldn't dare wear the same red and white painted stripes on their chests as the other Patriots if they wanted to operate discreetly, but their role in the movement was crucial.

Most recently many children were separated from their parents during the demonstration at Capitol Hill. News coverage of the incident claimed that a hundred or so rabble-rousers had been arrested, nobody was hurt, and everyone else calmed down and went home, but *street* coverage of the event said otherwise. Since the old-world independent news sources were deadlocked under the bureaucratic red tape, and anyone who challenged that mysteriously disappeared, the people of this once great nation would have to either swallow whatever lies the mainstream channels proselytized or die trying to dig for truth. Regional rumors abounded that there had been many deaths at Capitol Hill—maybe even more than that Georgetown riot, for crying out loud. But fake news aside, Danny was in the city the day that the Patriots marched on Capitol Hill, and he knew by the screams and the sirens that it was a bigger deal than the government wanted anyone to know about.

The White House's new administration wasn't responding as fast as the people wanted regarding the rescue of Earthquake Michaela victims still isolated on the Sacramento Islands—which is what folks had started calling the small remaining areas of California's capital city after it split in half. Danny was already in DC when that earthquake devastated one of the most prominent cities of the nation, but he *did* see the sun go black that day and later that same night the moon turn a million shades of evil, like blood. The ensuing meteor shower that fell on the Sacramento Islands the following morning took out a few thousand homes and began a ripple of enormous sinkholes. Mainstream news stations across the nation were covering every update until that point, showing heartbreaking footage of the wreckage, and more often independent newsrooms were interviewing famous religious personalities who made scary comments about God's sixth seal and Sacramento being swallowed by hell. Within a week almost every news station in the country had been shut down by the government, but not until after broadcasting a forced apology for "causing undue panic," "increasing vulnerability for national chaos," "contributing to acts of terrorism," and so on and so forth. What a load of horse hockey. This was, in Danny's opinion, the final straw that broke the camel's back on the freedom of speech. From that moment forward remaining stations continued to air, but to anyone with half a brain watching, it was obvious the stories were doctored. As far as the Sacramento Islands there was only a brief address from each news source that said the US president would release a statement soon.

That was months ago.

Of course everyone knew the United States didn't really have a president anymore. Sure, there was that skinny, sniveling Oval Office coward who loved to lick the boots of any suit within the New World Government Administration (NWGA), but nobody took his title as president seriously. It was clear he wasn't in charge and hadn't been since he was elected. But whether the absence of a true national leader on American soil was enough reason to explain the media blackouts on the obliteration of Sacramento or not, the people were still organizing marches in and around DC in an attempt to get answers about this fabled City of Refuge—and being stranded in this God-forsaken city at a time like this was becoming fatal.

Danny eyed the woman's food-filled purse again and proceeded cautiously. "OK, listen, listen..." His hands lifted to the air instinctively, palms outward in the universal gesture of surrender. "I'm not gonna take your food, OK? But I need your help. You gotta tell me how you got that."

She blinked and swallowed but said nothing.

"Because this guy," Danny continued, drawing her attention to the man behind the counter, whose expression was one of warning, "won't give me none of that." Danny pointed to her purse. "You know what I'm sayin', right? No food for me."

The ethnic diversity that he used to cherish about this tourist zone was going to be the death of him now. The translator app on his smartphone was no longer an option, as was the case for everyone else in the area and *maybe* the country at this point. For months the only way to communicate with anyone was by already knowing the same native language or enacting dramatic gestures. Ever since the morning that announcement came over the intercom at the Ronald Reagan Washington National Airport, alerting travelers that a severe weather phenomenon had grounded all unsanctioned flights indefinitely, all handheld digital communication devices were bugged out. Most of the time, the device screens still lit up, and many were finding that the flashlight icon worked, but everything else produced frozen glitch patterns. It didn't stop the public from trying, however. Danny had seen countless individuals cursing at their handheld devices, mashing their fingers on the app icons, and attempting to work their speech-to-text features as they normally would, but one by one, as the gadgets proved incapable of resurrection, widespread alarm launched violent outbreaks in the area that ended with the complete ransacking of nearby stores' electronics departments. Danny *would have* participated in all that if he thought stealing a brand-new phone would get him in contact with his family back home in Texas, but he knew well enough that an out-of-the-box device wouldn't work any better than all the other glorified flashlights if, as the people were saying, the NWGA officials were behind the global malfunction.

And of course they were. How could they *not* be? How else could the New World Government Administration keep people from reporting across state, region, and territory lines with additional tales of anarchy or natural disasters the second they happened? How else could they expect the people to eat up their claims that they were right on top of tragedies such as Earthquake Michaela and that there would be answers soon? *How else* could they spread their lies but render Americans in major cities powerless to communicate?

Danny wondered for a moment whether the post office behind the Gate still sent and received mail or if there was an excuse for that shutdown also. He would probably never know, since he wasn't allowed near the Gate to see for himself what sat beyond it. The stories he heard were nice, though...Government rations were better than the nothing he had.

Danny's stomach grumbled loudly. Agonizingly eternal seconds passed as he waited for the answer he just *knew* was coming from the woman with the food stash in her purse.

"No English."

Danny sighed and brought his hands to his hot cheeks. Blowing up wouldn't solve his hunger problem right now, and there was a very good chance—judging by the short fuse advertised on the restaurant owner's face—it *could* just get him shot. Closing his eyes momentarily, he heard the woman scuffle around him in a predictable escape, followed by the tinkling rattle of the small bell on the door of the main entry. He took a deep breath, plotted his next move in an instant, and opened his eyes to see the purse woman's frizzy mane dart around the corner. As calmly as possible, he collected his cash from the counter and stepped toward the door, hoping the restaurant owner would let him go without an issue. To his relief, the man appeared personally detached from caring about the woman's fate as Danny boldly pursued her. He must not have been a Patriot.

Bursting through the restaurant entrance and out into the pouring rain, Danny instinctively pulled his backpack straps, cinching them as tight as he could, and then threw his athletic football body into a full sprint after what he thought looked like yellow galoshes about twenty feet away. It was nearly impossible to see straight ahead, and even with the protection of his ball cap's bill, running with his eyes open meant inviting a massive shower of dust and debris from the wind into them. Yet the roar of the raindrops upon the pavement drowned out the dull thud of his sneakers behind her, and at least for the moment, the woman appeared as if she had no idea she was being followed.

Who would have known that this never-ending storm that had been plaguing the city for weeks would have been useful for something?

Ducking into an alley made the chase easier. Once he was sandwiched between two tall buildings, the overpowering down-pour was choked into a narrow slit, with the surrounding metal and brick absorbing most of the flow. The flood ten feet in slowed Danny down a bit, but it was about that time that the ball of frizz and yellow galoshes paused for a moment. Danny looked up and thought he saw the woman's face just before she dashed behind a dumpster.

"I'm not going to hurt you!" he shouted, decelerating his pace. "I just wanna ask you one question, and then I'll leave you alone, I promise! I'm not gonna take your food!"

There was no response, so he tried again as he grew closer.

"NO STEAL FOOD!...NO FIGHT!...OK? Just talk!"

Slowly, shakily, the woman poked her head out, exposing an expression of tremendous fright. Danny put his hands up and skirted a wide berth around her hiding place to face her more directly.

"Look man, I don't know if you all have some kind of DC-natives secret code around here, but I know you're doing *somethin'* to get you and your kids fed. You're a Patriot, right? One of them gals down on Rhode Island Avenue with all those kids? I got money, and it ain't no good anywhere in this city. I'm stranded here, and the Gate has been turning me away for months because I don't have the PalmChip. I can't prove I'm not related to any terrorists, so…"

Danny could tell she had no idea what he was saying, so he returned to exaggerated gestures: "Me," he said, pointing to his chest. "Danny…Danny no live here in DC." His arms made a giant arc above his head and then came down to pat his stomach. "Danny hungry. YOU get food from MAN."

His thumb was still pointing over his shoulder at the restaurant behind him when the woman put her finger up to her lips in a panic. "No sir, no sir. Please. Shhh."

"Whoa, OK, sorry," Danny said, realizing his loud mouth could expose this woman's food source. "How? *How* you get it?" When she shook her head in confusion, he gestured a gift exchange in the air between them, watching his volume. "What did you *give* him that he then *gives you* food? Money?" He pulled out the remaining eighty-dollar wad to show her. "You give money?"

The woman's face turned grave. "No sir."

Danny saw the smallest spark of understanding in her features and knew that if he was patient enough to stand in this river of an alleyway for just a few seconds more, he might get some answers, so he stared into her unrelentingly. Suddenly, with a deep breath and small shake of her head, the woman shifted her purse further up her arm, making a gun-shape with her hand and mimicked firing it.

"Bullet. Give bullet."

The familiar sounds of chaos and shouting on the street outside jerked Danny out of another pathetically superficial rest on the stink of stacked rubber mats.

"Hey, bro, you seein' this?" a voice echoed from the other side of the wall.

It had been possibly about twenty-four hours since Danny had left that probably Russian woman in the alleyway near the Mexican restaurant with the NO FOOD sign. Since that moment, he hadn't tried again to venture out for a meal. What was the point? He had nothing to offer either side now. Where his paper money was worthless, he had no ammunition currency or anything else of value. Where the food was free to the poor and cheap to wealthy behind the Gate—that nicer part of town where the streets were safely lined with security personnel and people still operated with some form of commerce—he was viewed as a potential terrorist. Anyone who rejected the PalmChip must have something to hide.

Maybe his homeys back in Missouri had been onto something when they told him he should receive the optional government-issued ID implant, PalmChip, which stored all your identification information, debit card, credit card, and medical history "in the palm of your hand." Of course this marketing tagline was ironic, because everyone knew the implant was actually inserted in the *back* of the hand, but that didn't have as impressive a ring to it.

With starvation as a reality looming ahead of him now, the chip didn't seem nearly as optional. It was literally becoming the reason he couldn't beg, borrow, steal…or *trade* for a meal.

The thin wall rattled and pounded as the voice hollered out

again. "'Ey, Danny, you awake? You seein' this? Man, look out your window!"

Danny, who was sacked out on the floor, staring in the opposite direction of the window, ignored his buddy and went back to the one thought that continued to haunt him all hours of the day: in a way, though the details were different, this chip thing was a little too similar to what his Gramma Cadence told him would go down with that triple-six tattoo everyone was supposedly going to get right before the end of the world.

Some ten years earlier, when Danny was around the age of sixteen, Gramma Cadence had stirred an enormous stink amid their religious friends in their hometown area when she stood in the grocery department of a Walmart scream-debating the local minister about a woman's right to preach and teach in a church. Since that seriously crazy display the stamp that had been imprinted in Danny's mind was that his grandmother was off her rocker. He loved her, and he missed her every day since her untimely death just after his eighteenth birthday, but whenever he thought of her as a biblical authority, his mind would flash back to that Walmart humiliation, and all he could see was her ranting and raving and throwing apples.

Now, stuck in DC without any direction, without any home besides this fire-damaged and flooded old flower shop that smelled of urine and mold, Danny would give anything to talk to her just once more about this "man of sin" she used to talk so much about. That he knew of, nobody was getting a silly "666" tattoo, but wouldn't that have been too obvious anyway?

Then again, Danny was starving, and the PalmChip was, according to the armed men at the Gate, still a viable option for him if he would just—

"Man, did you die on me or somethin'?" the voice called once more as another round of fist-beatings rattled the almost-collapsing wall. "Danny, man, look out your window!"

"Calm down, Mike," Danny said, curling himself to his side. "I'm trying to sleep, and whatever it is doesn't concern me anymore."

From the giant charred hole in the wall at the end of the room, another tall African American male stepped into Danny's space, ignoring the glass-crunch sound as he walked past Danny and over to the window. Mike had been scheduled to be on Danny's same flight to Chicago the day the flights were grounded and the cell phones started bugging out. The two young men had struck up a conversation outside the boarding terminal when Mike almost picked up Danny's identical backpack, mistaking it for his own, and they were still talking intermittently through three reported delays. After Danny's cell phone started beeping a bizarre error signal and shut off by itself, the final *weather phenomenon* announcement hit the intercom amid a wave of gasps and tears from heartbroken travelers. Mike and Danny shared a knowing look, and Mike hurriedly dug for his own phone, which powered up normally and appeared to navigate just fine, but no calls or texts would go through. Within the next few minutes both of the boys were approached by panicked passersby asking if they could borrow a phone since their own wasn't getting reception or had shut off altogether. Laptops were kicked off the airport's public Wi-Fi, the staff at the terminal fiddled about trying to get the televisions to tune in to anything other than static, and when they were successful to tap into the one channel with a strong signal, it was NWGA News.

From that surreal moment to this one, Danny and Mike were inseparable. As luck would have it, Mike's family members were religious freaks also, so he had refused the PalmChip just as Danny had, and the stiff-necked Gatekeepers wouldn't budge on the food distribution.

"Dude, it's the Patriots, and they got guns!" Mike said, returning to Danny's side and pulling his arm. "Bro, this is something serious. It's happening down at the parking garage. With or without you, I'm going to check it out."

Danny, bewildered, rose to his feet and finally ran to check the window. "This doesn't make sense, Mike! What are the Patriots doing invading the parking garage? Surely they know there are hundreds of harmless families holed up inside...What do they want with *them*?"

"I don't think they're invading, dude. Look at 'em. They're protecting something in the middle there...They're bringing something in! Do you think they got food?"

Without another word, Danny found a final string of energy to bound from his place in the moldy flower shop, followed on his heels by Mike. Both of them grabbed their backpacks and swung them on, though neither of them bothered to grab their piles of old-world cash, which had now proved to be absolutely worthless. Danny always slept in his sneakers, never knowing when he may need to suddenly book it to a new crash pad in the middle of the night, so he was ready to go, and he ran like the wind down the soaked, cracked sidewalks. This past week, since they found the empty flower shop, Danny and Mike both had to tread carefully around these blocks and try to keep distance between them and any other hungry squatters with mugging on the mind. But this time they moved with complete abandon, undeterred by the facial blast of rainwater each time they found themselves between awnings.

The nooks and crannies that were usually filled with small groups in a huddle were vacant, and up ahead others were pouring out of nearby buildings to join the crowd near the Patriots. As Mike and Danny got closer, they could hear a sudden hush fall over the gathering. One man had elevated above the rest, obviously standing on some kind of platform. He was making a loud announcement, but between the sound of the hard rain and the heavy breathing from a sudden jolt of exercise, they couldn't make out a word of it. Right away the crowds started to form into long, single-file lines. The man stepped back down, fading again into the sea of heads.

As the lines of people stretched further out from the building, they met up with Danny and Mike to cover the last block to the garage. Amid the diverse ethnicities one young girl at the very back appeared approachable as she smiled their way.

"Exciting, isn't it?" she shouted with a jump and a clap. "We're getting food!"

"Everyone? Even the street people?" Danny yelled back, trying to be heard above the rain and the crowd, which had resumed their chatter. "Why? How?"

"The Patriots have some guy on the inside. They organized a midnight raid down at Gate Seven several days ago, and they're only just making it to our part of town."

Danny caught Mike in his peripheral vision making prayer hands beneath his chin. He thought for a moment he also saw Mike release a sob.

"Wait," Danny said. "I'm confused... *We* aren't part of the Patriot demonstrations. We've never put *our* lives in danger to make demands. Why would they share their food with us when we haven't done anything to earn it?"

"Maybe because they're decent and we're *people*?" the girl barked with a shrug. "Maybe they don't agree with the NWGA that we should have to starve just because we won't be chipped. Or maybe they're recruiting. I don't know. Nor do I care. I'm starving!"

Danny stood on tiptoe, arching his head as far past the gathering as he could, trying to catch a glimpse of the entrance to the garage. Only for a moment a clear path opened and revealed that at the front of the line the Patriots were operating scanners.

"Mike! Scanners! I saw scanners! What the..."

"I have no idea, bro," Mike said from behind bloodshot, emotional eyes. "They're smart enough to know by now that none of us are chipped, else we'd already be livin' behind the Gate. But whatever, man. What do we have to hide? Let's just get scanned and see what this free-food thing is all about."

Danny blinked helplessly as his buddy once again made prayer hands and, despite the downpour, turned his face to the sky as if in prayer. Danny hoped for Mike's sake that this too-good-to-be-true food lineup was what it looked like.

The line was moving quickly and steadily, which wasn't a surprise. All the people in this area had to have been eager to cooperate if it meant getting their hands on rations. Danny kept his eyes curiously ahead at the approaching garage entrance, which had what looked like a good hundred or so armed Patriot soldiers lined up and more than ready to shoot anyone who didn't act in an orderly way. His mind raced through every possibility to explain what they were walking into, but in the end it didn't matter. What choice did he have? Dying of starvation on the floor of the flower shop?

A few Patriots were walking out into the lines with their hands up, shouting, but none of it was in English.

"What are they saying?" Danny shouted at the group ahead of him. "Does anyone understand what they're talking about?"

A Latina woman turned to offer an answer. "They're asking for translators." Turning back toward the front, she raised her hand and flagged the soldier down, rattling off something in Spanish. He nodded and gestured for her to follow him to the front, and after a brief exchange he nodded again, and she turned to collect what Danny assumed was her family before walking away. Others throughout the lines were being pulled out as well, each of them holding quick conversations in other languages and then heading to the front of the line.

After what felt like an eternity, Danny and Mike arrived at the garage. Mike put both his hands out, palm side down, like the last several hundred people before him had done, and a Patriot passed a long red light like a glowing nightstick over them. There was a buzzing noise and a flash of red, and then the soldier smiled at Mike and gestured him in. It went down identically for Danny, who uneasily stepped forward, half expecting to be abducted around every corner. Just on the other side of the entrance were soldiers holding their guns tight and herding people in different directions.

"English speakers, red stairwell, stop on the second floor!" one of them was yelling over and over.

Mike and Danny trailed along with the masses toward one of the six sets of stairs. The color red had been spray-painted above and around it to help guide everyone. The elevators weren't responding, so people were bottlenecking into the narrow stairwell. Thankfully, Mike and Danny only had to get one floor up.

"Keep moving; food is just around the corner," a Patriot said, smiling at the hordes of people who were growing increasingly emotional. All around him men and women of all ages were pelting him with questions.

"Is this for real?"

"Is there really food up here?"

"Is this a trap?"

The Patriot laughed. "I assure you folks, it's all legit, as you will see in just a moment. Right this way; keep walking."

When Danny's sneakers grabbed the edge of the final stair to the second floor, he almost lost his cool for a moment. There, where hundreds of useless old-world cars had been turned into family shelters, Patriots were passing out protein bars, bags of popped corn, tortilla chips, cereal...more food than Danny had seen in months, since he was last at the airport. Mike was openly crying now.

"See, I told you, Danny. I told you, man...We're gonna live! We're gonna survive this thing, man!"

Danny might have responded, but right at that moment a Patriot rounded the end of a row of cars and spun in Danny's direction. With a smile he handed them each a small bag of rations with a bottle of water at the bottom, and Danny wasted no time tearing into them. Both he and Mike ravaged the bag like famished wolves on a pile of steaks.

For about a full minute, neither one said a thing while they choked the food down, and before they had a chance to say another word, the next set of instructions was coming from the Patriots who were now grouping people as tightly in around the car shelters as possible.

"Squeeze in, squeeze in. Close in this row over here, guys. Good, good. Get in as tight as possible. He'll be here in a sec."

"Who?" everyone was shouting at once.

"The leader of the Patriots. He's starting with the English floor, so you need to be ready. He will explain everything. He's on his way now."

For the next couple of minutes the floor grew relatively quiet, except for a nearby radio playing the latest round of NWGA News jargon. Most of these people hadn't eaten in so long that they weren't interested in visiting as much as chewing, and even though everyone spoke English, the free-food scene was so unbelievable that there was a corporate speechlessness hanging thick in the air. So when a sudden burst of chaos erupted at the end of the rows near the stairwell and people all over the place were jumping to their feet and pointing, Danny and Mike shared a nervous glance and a deep breath. Instinctively they both crumpled their food into their backpacks and stood with everyone else, preparing to crowd the entryway, when a voice came over a megaphone instructing folks to move aside so the leader of the Patriots could make it through.

When the throng dispersed again, a masked figure dressed completely in black walked to the end of the row of cars, surrounded by a group of soldiers. The soldier with the megaphone stepped to the front of their assembly.

"Ladies and gentlemen, before our leader reveals what he has to say, you have to understand that it is *crucial* you remain calm. This man has only minutes to tell you and every foreign-language floor above you what he has to say, and I assure you, it's a matter of *more* than life or death that you listen. Hold all questions until he's gone, and a Patriot will be assigned to your groups to respond to any concerns you have. Do not—I repeat, *do not*—attempt to follow, run after, physically approach, or delay this individual at any point. His guards are ready and willing to use fatal force if this advice is not heeded."

From there the soldier passed the megaphone to the masked man. As he took it in one hand, he reached for his mask in the other and twisted it off in one graceful movement. Danny didn't immediately recognize the guy, but he marveled silently at how young he was.

"The leader's a *kid*?" Danny whispered.

"Quiet, please!" the leader said nervously. "Before I get into what I have to say, I just want you all to know one thing: The reason you all have food is because Amelia McCutcheon organized underground pantry distribution programs before she and Pete disappeared. Her work continues on under the guise Amelia's Angels, and those blessed women are the saints behind the Gate Seven raid a few days back. Now, let me get to why I'm here now and what I have to say to all of you."

Danny could tell by the expressions of those around him that he wasn't alone in thinking that shift in the leader's speech was abrupt, but he was about to find out why.

"I don't have time for a grand introduction or any getting-to-know-you exchanges, as I fully expect to be arrested any minute. I'm going to skip straight to the point. My name is Ben Brandeis. You don't know me, but our government does. I'm a wanted man. For the past three years I have been part of a classified group at the highest level of the White House. Pete and Amelia McCutcheon were direct associates of mine. I worked with them daily. Our top-secret mission had involved efforts to deflect Asteroid 2004 JU04, nicknamed Wormwood, from colliding with Earth this year."

Some found that assertion shocking, and there was a stir of exchanged glances, but in that moment, with full bellies for the first time in a long while, any insane claim sounded plausible. Ben continued, speaking almost so quickly that it was easy to get lost a sentence or two behind. Danny forced his swirling thoughts to the back of his mind, determined to take it all in.

"I went rogue several months back when the PalmChip became mandatory for any US official. I had only heard of the Patriots and their work up until that point, but when I had to go into hiding, I didn't have anywhere else to go."

Ben nodded graciously toward his guards.

"They took me in, thank God, and as it turns out, they needed my intel as much as I needed their protection and support. I didn't take the chip, and according to our scanners at the entrance of this garage, neither did any of you. You've done the right thing. Every one of you has been plagued by a gut feeling that said this chipping system was exactly what the religious fanatics are saying it is, that we're in the end times as described in the Book of Revelation, and that the man our good-for-nothing president constantly worships on TV is the Antichrist. If you've paid as close attention, as I imagine most of you have, to the suspiciously unrealistic news loops, then you'll already know that our government leaders left us *months* ago and stationed themselves underground in the Refuge Cities—which are not, as that moron Steve Campion says, fictional. I guarantee you they are very real, and before you interrupt me,

I guarantee you they are very full and any unauthorized person within ten miles of these Refuge City tunnels will be shot on sight."

The people listened without moving an inch. While Ben paused, the only sound that could be heard was the din of the floors above, still settling in with their ration bags in different languages happily, still having no clue about what they were about to hear.

"Folks," Ben continued, "I wish there was a way to soften the blow, but hard and fast is the only way to tell you. Wormwood was not the only space rock our radars picked up; it was just the largest threat, which was why the US put the best men on that team. There was another, asteroid 2016 BQ17, that hit the Gulf of Mexico, causing a tsunami that wiped out the power grid throughout most of Texas and Northern Mexico. Although damage assessment is nearly impossible while our government is absent, inside sources tell me that Texas is obliterated."

"Oh God, no," Danny whispered. Though he wasn't into religion before, he found himself very willing to communicate to God.

"Please God, not my family. Not my sister. Please no. Say they're OK…"

"This is information you will not hear on the news until the NWGA finds a way of spinning it to make their main man look like a hero," Ben said, closing his eyes for a moment before continuing. "Lastly, all efforts to redirect Wormwood have failed. It's the size of Africa, with six orbiting companion asteroids, but what you need to know is that they're calling it a planet killer, and it's on a trajectory to hit in thirty-six to forty-eight hours from now. Why am I telling you all this? It's not to make you sad. I gotta get real with you because you have a very serious, very *eternal* choice to make right now. There are two ways to react at this point. Panic and see how many people can be foolishly and pointlessly trampled to death, or get on your knees and pray. God *is* in control, and because you haven't been chipped, you each have the opportunity to give yourselves over to *THE—ONLY—ONE* who can save you now..."

Ben took a deep breath, shook his head, swallowed, and blinked, displaying for the first time his own emotional turmoil.

"Pete McCutcheon tried to tell me all of this. He tried. But it was too much to take in at the time. All of this—from the destruction of Sacramento forward—McCutcheon told me exactly how it would happen, and he was right..."

Danny was too shocked to cry. As Ben's words drifted into unintelligible mumbling in the background, Danny glanced over at Mike, who had fallen to his knees. Danny could faintly hear him reciting that one scripture about a valley of death. A quiet hum of weeping had started to grow here and there amid the people when Danny tuned back in.

"We expect impact of this asteroid by Friday afternoon. If you would be so inclined to trust in the Lord at this time, I invite you to pray with me now..."

NOTES

AUTHOR'S NOTE

1. "Asteroids," NASA Science, accessed July 23, 2019, https://solarsystem.nasa.gov/asteroids-comets-and-meteors/asteroids/in-depth/.
2. "Asteroid or Meteor: What's the Difference?" NASA Science, updated July 17, 2019, https://spaceplace.nasa.gov/asteroid-or-meteor/en/.
3. "Asteroid or Meteor: What's the Difference?" NASA Science.
4. "Asteroid or Meteor: What's the Difference?" NASA Science.
5. "Asteroid or Meteor: What's the Difference?" NASA Science.

CHAPTER 2
EYES ON THE SKIES

1. Alex Renton, "Nathan Myhrvold, Myth Buster," *The Economist*, January/February 2015, https://www.1843magazine.com/content/features/myth-buster.
2. Nathan Myhrvold, "An Empirical Examination of WISE/NEOWISE Asteroid Analysis and Results," *Icarus* 314 (November 1, 2018), 64–97.
3. Nathan Myhrvold, "What's Wrong With NEOWISE," Medium, June 14, 2018, https://medium.com/@nathanpmyhrvold/myhrvold-guide-to-neowise-4866a2f7b76d.
4. Myhrvold, "What's Wrong With NEOWISE."
5. Myhrvold, "An Empirical Examination of WISE/NEOWISE Asteroid Analysis and Results."
6. Harry Lear, "Open Letter to President Trump," *They Fly Blog*, January 29, 2018, https://theyflyblog.com/2018/02/15/open-letter-to-president-trump/.
7. Lear, "Open Letter to President Trump."

8. Sean Martin, "NASA Warning: Terrifying Advice for Life-Ending Asteroid Strike on Earth," *Express*, July 4, 2019, https://www.express.co.uk/news/science/1149052/space-nasa-news-earth-asteroid-2019-asteroid-belt-meaning-hit-earth.

9. Martin, "NASA Warning."

10. "ESA Confirms That Asteroid Will Miss Earth in 2019," European Space Agency, accessed July 29, 2019, http://www.esa.int/Our_Activities/Space_Safety/ESA_confirms_asteroid_will_miss_Earth_in_2019.

11. Kelsey Piper, "A 'City-Killing' Asteroid Just Zipped by Earth. Why Didn't We See It Coming?," Vox, July 26, 2019, https://www.vox.com/future-perfect/2019/7/26/8931776/near-earth-asteroid-tracking.

12. Institute for Astronomy at the University of Hawaii, "Breakthrough: UH Team Successfully Locates Incoming Asteroid," news release, June 25, 2019, http://www.ifa.hawaii.edu/info/press-releases/ATLAS_2019MO/?fbclid=IwAR2Dumlqr3E2PPL70Z2h3nV l8dzfzZoDw96egq__L60GPn0GboMK_czkpck; Marshall Shepherd, "Scientists Detected an Incoming Asteroid the Size of a Car Last Week—Why That Matters to Us," *Forbes*, June 27, 2019, https://www.forbes.com/sites/marshallshepherd/2019/06/27/scientists-detected-an-incoming-asteroid-the-size-of-a-car-last-week-why-that-matters-to-us/#e8234a848698.

13. Scott Carney, "Did a Comet Cause the Great Flood?," *Discover*, November 15, 2007, http://discovermagazine.com/2007/nov/did-a-comet-cause-the-great-flood.

14. Whitney Clavin and J. D. Harrington, "Herschel Spacecraft Eyes Asteroid Apophis," NASA, January 9, 2013, https://www.nasa.gov/mission_pages/herschel/news/herschel20130109.html.

15. Greg Bear, "Don't Count 'Doomsday Asteroid' Out Yet," CNN, January 24, 2013, https://www.cnn.com/2013/01/23/opinion/bear-apophis-asteroid/index.html.

16. Samantha Mathewson, "MIT Students Design Mission to Huge Asteroid Apophis," Space.com, June 7, 2017, https://www.space.com/37117-mit-students-design-asteroid-apophis-mission.html.

17. Artie Villasanta, "Scientists Dismiss Reports a 200 Meter-Wide Asteroid Will Strike the Earth in 2023," *Business Times*, November 27, 2018, https://en.businesstimes.cn/articles/105286/20181127/scientists-dismiss-reports-200-meter-wide-asteroid-will-strike-earth.htm.

18. Mathewson, "MIT Students Design Mission to Huge Asteroid Apophis."

19. David Noland, "5 Plans to Head Off the Apophis Killer Asteroid," *Popular Mechanics*, November 7, 2006, https://www.popularmechanics.com/space/deep-space/a1025/4201569/.

20. Bear, "Don't Count 'Doomsday Asteroid' Out Yet."
21. Bear, "Don't Count 'Doomsday Asteroid' Out Yet."
22. Bear, "Don't Count 'Doomsday Asteroid' Out Yet."
23. Carney, "Did a Comet Cause the Great Flood?"
24. Alexandra Lozovschi, "Large Asteroid Packing 50 Megatons of Force Might Come Crashing Down on Earth in 2023—and That's Not All," Inquisitr, November 25, 2018, https://finance.yahoo.com/news/large-asteroid-packing-50-megatons-214202306.html.
25. Sebastian Kettley, "NASA Asteroid WARNING: 700-Foot-Wide Space Rock on 62 RISK Trajectories With Earth by 2023," *Express*, updated November 28, 2018, https://www.express.co.uk/news/science/1050061/NASA-asteroid-warning-Earth-collision-risk-2023-Asteroid-LF16.
26. Villasanta, "Scientists Dismiss Reports a 200 Meter-Wide Asteroid Will Strike the Earth in 2023."
27. "Asteroid as Powerful as 50 Megaton Nuke May Slam Into Earth in 2023—NASA," Physics-Astronomy, November 17, 2018, http://www.physics-astronomy.com/2018/11/asteroid-as-powerful-as-50-megaton-nuke.html#.XTCgcuhKiiM.
28. Sebastian Kettley, "NASA Asteroid Tracker: A 1,115FT Asteroid Skimmed the Earth TODAY at 34,000MPH," *Express*, March 27, 2019, https://www.express.co.uk/news/science/1105076/NASA-asteroid-tracker-Asteroid-2019-EN-skim-earth-close-approach; "JPL Small-Body Database Browser: 2019 EN," NASA, accessed September 22, 2019, https://ssd.jpl.nasa.gov/sbdb.cgi?sstr=2019%20EN;old=0;orb=0;cov=o;log=0;cad-1#cad.
29. "JPL Small-Body Database Browser," NASA.
30. Inigo Monzon, "This New Asteroid Could Destroy Earth in 50 Years, Korean Scientists Reveal," *International Business Times*, June 27, 2019, https://www.ibtimes.com/new-asteroid-could-destroy-earth-50-years-korean-scientists-reveal-2803318.
31. "Risk List," European Space Agency, accessed September 22, 2019, http://neo.ssa.esa.int/risk-page.
32. "Four Asteroids on Collision Course With Earth," RT, June 30, 2019, https://www.rt.com/news/463071-asteroids-collision-course-earth/.
33. "Four Asteroids on Collision Course With Earth," RT.
34. "Risk List," European Space Agency; "JPL Small-Body Database Browser: 2000 SG344," NASA, accessed September 22, 2019, https://ssd.jpl.nasa.gov/sbdb.cgi?ID=bK00SY4G;old=0;orb=0;cov=0;log=0;cad=1#cad.
35. Media Gazelle, "A 2,455 Foot Wide Asteroid Is Hurtling Towards Earth Today," *US Tribune News Online*, April 6, 2019, https://ustribune.news/2019/04/06/a-2455-foot-wide-asteroid-is-hurtling-towards-earth-today-2/.

36. Sebatstian Kettley, "Asteroid Bennu Danger: Will Asteroid Bennu Hit Earth? What Are NASA's Odds of Doom?," *Express*, January 31, 2019, https://www.express.co.uk/news/science/1080895/Asteroid-bennu-nasa-danger-will-asteroid-bennu-hit-earth.

37. Mike Wall, "No, Asteroid Bennu Won't Destroy Earth," Space.com, August 1, 2016, https://www.space.com/33616-asteroid-bennu-will-not-destroy-earth.html.

38. Afiya Qureshi, "Three Rogue Asteroids May Collide With Earth This Century," *Mashable*, accessed July 3, 2019, https://in.mashable.com/science/4514/three-rogue-asteroids-may-collide-with-earth-this-century.

39. "About OSIRIS-REx," NASA, accessed August 5, 2019, https://www.nasa.gov/mission_pages/osiris-rex/about.

40. "101955 Bennu," NASA, accessed August 5, 2019, https://solarsystem.nasa.gov/asteroids-comets-and-meteors/asteroids/101955-bennu/overview/.

41. "Where Is the Spacecraft?" Arizona Board of Regents, accessed July 5, 2019, https://www.asteroidmission.org/where-is-the-spacecraft/.

42. Lonnie Shekhtman, "Why Bennu? 10 Reasons," NASA, August 20, 2018, https://solarsystem.nasa.gov/news/517/why-bennu-10-reasons/.

43. Shekhtman, "Why Bennu?"

44. "Double Asteroid Redirection Test (DART) Mission," NASA, accessed September 22, 2019, https://nasa.gov/planetarydefense/dart; Tricia Talbert, ed., "NASA's First Planetary Defense Technology Demonstration to Collide With Asteroid in 2022," NASA, May 6, 2019, http://nasa.gov/feature/nasa-s-first-planetary-defense-technology-demonstration-to-collide-with-asteroid-in-2022.

45. Benjamin Raven, "NASA's Planetary Defense Test Set to Launch in 2021, Collide With Asteroid in 2022," MLive Media Group, May 13, 2019, https://www.mlive.com/news/2019/05/nasas-planetary-defense-test-set-to-launch-in-2021-collide-with-asteroid-in-2022.html.

46. "Breaking Up Is Hard to Do: Asteroids Are Stronger, Harder to Destroy Than Previously Thought," Johns Hopkins University, March 4, 2019, https://releases.jhu.edu/2019/03/04/breaking-up-is-hard-to-do-asteroids-are-stronger-harder-to-destroy-than-previously-thought/.

47. "Asteroid Fast Facts," NASA, accessed August 5, 2019, https://www.nasa.gov/mission_pages/asteroids/overview/fastfacts.html.

48. "Asteroids," NASA Science.

49. Raven, "NASA's Planetary Defense Test Set to Launch in 2021, Collide With Asteroid in 2022."

50. Ryan Whitwam, "SpaceX Wins Contract to Launch NASA's DART Asteroid Impactor," ExtremeTech, April 15, 2019, https://www

.extremetech.com/extreme/289518-spacex-wins-contract-to-launch
-nasas-dart-asteroid-impactor.

51. Raven, "NASA's Planetary Defense Test Set to Launch in 2021,
Collide With Asteroid in 2022."

52. Matthew Knight, "New Confirmation of Billy Meier's Warnings
About Apophis," *They Fly Blog*, May 6, 2015, https://theyflyblog
.com/2015/05/04/new-confirmation-of-billy-meiers-warnings-about
-apophis/.

53. Lear, "Open Letter to President Trump."

54. Lear, "Open Letter to President Trump."

55. "Episode 49: Billy Meier, Michael Horn, and Asteroid Apophis,"
Exposing PseudoAstronomy Podcast Shownotes, accessed August 5,
2019, https://podcast.sjrdesign.net/shownotes_049.php.

56. Thomas Horn, *Saboteurs* (Crane, MO: Defender Publishing, 2017),
82–83.

57. Lear, "Open Letter to President Trump."

58. Lear, "Open Letter to President Trump."

59. Melissa Nord, "'We Did All We Could, but It Wasn't Enough' |
Asteroid Simulation Ends With Direct Hit in NYC," WUSA-TV,
updated May 10, 2019, https://www.wusa9.com/article/weather/
weather-blog/we-did-all-we-could-but-it-wasnt-enough-asteroid-
simulation-ends-with-direct-hit-in-nyc/65-f63d4b6b-3314-4026-
a005-affc3d7dd121.

60. Nord, "'We Did All We Could, but It Wasn't Enough.'"

61. Nirmal Narayanan, "A Doomsday Asteroid Could Inevitably Hit
Earth, Predicts Top British Scientist," *International Business Times*,
India Edition, November 16, 2018, https://www.ibtimes.co.in/
doomsday-asteroid-could-inevitably-hit-earth-predicts-top-british-
scientist-785704.

62. Narayanan, "A Doomsday Asteroid Could Inevitably Hit Earth,
Predicts Top British Scientist."

63. Solar System Exploration Research Virtual Institute, "6th IAA
Planetary Defense Conference—The Honorable James Bridenstine,
NASA Administrator," YouTube, April 29, 2019, https://www.
youtube.com/watch?v=JBv-FCZoScc&feature=youtu.be.

64. Solar System Exploration Research Virtual Institute, "6th IAA
Planetary Defense Conference."

65. Solar System Exploration Research Virtual Institute, "6th IAA
Planetary Defense Conference."

66. Deborah Byrd, "Preparing for Asteroid Apophis," EarthSky, April
30, 2019, https://earthsky.org/space/preparing-asteroid-apophis-
april-13-2029-passage.

67. Melissa Hogenboom, "In Siberia in 1908, a Huge Explosion Came
Out of Nowhere," BBC, July 7, 2016, https://www.bbc.com/earth

/story/20160706-in-siberia-in-1908-a-huge-explosion-came-out-of
-nowhere.

68. Hogenboom, "In Siberia in 1908, a Huge Explosion Came Out of
Nowhere."
69. Hogenboom, "In Siberia in 1908, a Huge Explosion Came Out of
Nowhere."
70. Hogenboom, "In Siberia in 1908, a Huge Explosion Came Out of
Nowhere."
71. Hogenboom, "In Siberia in 1908, a Huge Explosion Came Out of
Nowhere."
72. Hogenboom, "In Siberia in 1908, a Huge Explosion Came Out of
Nowhere."
73. Villasanta, "Scientists Dismiss Reports a 200 Meter-Wide Asteroid
Will Strike the Earth in 2023."
74. Anurag Ghosh, "What Is the Average Asteroid Speed?," Bright Hub,
accessed August 6, 2019, https://www.brighthub.com/science/space
/articles/64710.aspx.
75. Aristos Georgiou, "God of Chaos Asteroid Apophis Is Headed
for Earth—and NASA Is Excited," *Newsweek*, April 30, 2019,
https://www.newsweek.com/god-chaos-apophis-earth-nasa-
excited-1409509.
76. Joshua J. Mark, "Apophis," *Ancient History Encyclopedia*, April 25,
2017, https://www.ancient.eu/Apophis/.
77. Brian Handwerk and John Roach, "Where Our Fear of Friday the
13th Came From," National Geographic Society, November 13,
2015, https://news.nationalgeographic.com/2015/11/151113-friday-
13-superstition-phobia-triskaidekaphobia-culture/.
78. Thomas Horn, "Saboteurs Part 19: Shadow Government and
Magic Numerology," SkyWatch TV, October 5, 2017, https://www.
skywatchtv.com/2017/10/05/saboteurs-part-19-shadow-government-
magic-numerology/.
79. Ethelbert W. Bullinger, *Number in Scripture* (Grand Rapids, MI:
Kregel Publication 2003), 205.

CHAPTER 3
OUR DESTINY BEYOND EARTH

1. Galileo Galilei, "Galileo Galilei Quotes," BrainyQuote, accessed
August 6, 2019, https://www.brainyquote.com/quotes/galileo
_galilei_136976.
2. Matt Williams, "Who Was Galileo Galilei?" Universe Today,
November 5, 2015, https://www.universetoday.com/48756/galileo-
facts/.

3. Justin Bachman and Travis Tritten, "Why Trump Wants a Space Force for the Final Frontier," *Washington Post*, February 19, 2019, https://www.washingtonpost.com/business/why-trump-wants-a-space-force-for-the-final-frontier/2019/02/19/aac0b1ee-349d-11e9-8375-e3dcf6b68558_story.html?utm_term=.2dfd0b3236da.

4. Valerie Insinna, "Trump Officially Organizes the Space Force Under the Air Force…for Now," Defense News, February 19, 2019, https://www.defensenews.com/space/2019/02/19/trump-signs-off-on-organizing-the-space-force-under-the-air-forcefor-now/.

5. Bachman and Tritten, "Why Trump Wants a Space Force for the Final Frontier."

6. Insinna, "Trump Officially Organizes the Space Force Under the Air Force…for Now."

7. Bachman and Tritten, "Why Trump Wants a Space Force for the Final Frontier."

8. Bachman and Tritten, "Why Trump Wants a Space Force for the Final Frontier."

9. Bachman and Tritten, "Why Trump Wants a Space Force for the Final Frontier.".

10. Insinna, "Trump Officially Organizes the Space Force Under the Air Force…for Now."

11. Insinna, "Trump Officially Organizes the Space Force Under the Air Force…for Now."

12. Eric Mack, "Trump's Space Force: Everything You Need to Know," CNET, January 16, 2019, https://www.cnet.com/news/trump-space-force-everything-you-need-to-know/.

13. Michael Sheetz and Amanda Macias, "Trump Officially Directs Pentagon to Create Space Force Legislation for Congress," CNBC, February 19, 2019, https://www.cnbc.com/2019/02/19/trump-directs-pentagon-to-create-space-force-legislation-for-congress.html.

14. Sheetz and Macias, "Trump Officially Directs Pentagon to Create Space Force Legislation for Congress."

15. Sheetz and Macias, "Trump Officially Directs Pentagon to Create Space Force Legislation for Congress."

16. Bryan Bender, "National Security Leaders Urge Congress to Back Space Force," POLITICO LLC, May 6, 2019, https://www.politico.com/story/2019/05/06/space-force-congress-1413691.

17. Mack, "Trump's Space Force."

18. "X-37B Orbital Test Vehicle," US Air Force, September 1, 2018, https://www.af.mil/About-Us/Fact-Sheets/Display/Article/104539/x-37b-orbital-test-vehicle/.

19. Eric Mack, "SpaceX Launches Hush-Hush Space Plane," *Cnet.com*. September 7, 2019, https://www.cnet.com/news/elon-musk-spacex-launches-air-force-x37-b-space-plane/.

20. Leonard David, "X-37B Military Space Plane Wings Past 400 Days on Latest Mystery Mission," *Space.com*, October 18, 2018, https://www.space.com/42175-x-37b-space-plane-otv5-400-days-orbit.html.

21. "Air Force Research Laboratory—Space Vehicles Directorate: Advanced Structrually Embedded Thermal Spreader II (ASETS-II)," US Air Force, July 2017, https://www.kirtland.af.mil/Portals/52/documents/ASETSII.pdf?ver=2017-08-10-125001-450

22. CNN, "Trump Directs Formation of a Space Force," YouTube, June 18, 2018, https://www.youtube.com/watch?v=CAM9gvH7gFE.

23. "Treaty on Principles Governing the Activities of States in the Exploration and Use of Outer Space, Including the Moon and Other Celestial Bodies," US State Department, accessed August 7, 2019. https://2009-2017.state.gov/t/isn/5181.htm.

24. "Treaty on Principles Governing the Activities of States in the Exploration and Use of Outer Space, Including the Moon and Other Celestial Bodies," US State Department.

25. Loren Grush, "How an International Treaty Signed 50 Years Ago Became the Backbone For Space Law," The Verge, January 27, 2017, https://www.theverge.com/2017/1/27/14398492/outer-space-treaty-50-anniversary-exploration-guidelines.

26. "Outer Space Treaty of 1967," NASA, accessed August 8, 2019, https://history.nasa.gov/1967treaty.html.

27. Mack, "Trump's Space Force."

28. Bachman and Tritten, "Why Trump Wants a Space Force for the Final Frontier."

29. "WISE Mission Overview," NASA, accessed August 8, 2019, https://www.nasa.gov/mission_pages/WISE/mission/index.html.

30. "Mission Overview," NASA, accessed August 8, 2019. https://www.nasa.gov/mission_pages/neowise/mission/index.html.

31. "Large Synoptic Survey Telescope," LSST, accessed August 8, 2019, https://www.lsst.org/.

32. "LSST," Association of Universities for Research in Astronomy, accessed August 8, 2019, https://www.aura-astronomy.org/centers/large-synoptic-survey-telescope/.

33. Loukia Papadopoulos, "NASA Is Working on a Camera to Save Humanity," Interesting Engineering, April 22, 2019, https://interestingengineering.com/nasa-is-working-on-a-camera-to-save-humanity.

34. "LSST," Association of Universities for Research in Astronomy.

35. "Large Synoptic Survey Telescope," LSST.

36. "Software Success," LSST, accessed August 8, 2019, https://www.lsst.org/news/software-success.

37. Papadopoulos, "NASA Is Working on a Camera to Save Humanity."

38. "NASA's First Asteroid Deflection Mission Enters Next Design Phase," NASA, June 30, 2017, https://www.nasa.gov/feature/nasa-s-first-asteroid-deflection-mission-enters-next-design-phase.

39. "NASA's First Asteroid Deflection Mission Enters Next Design Phase," NASA.

40. Tariq Malik, "How Trump's Space Force Would Help Protect Earth From Future Asteroid Threats," Future US, Inc., June 20, 2018, https://www.space.com/40949-trump-space-force-asteroid-defense.html.

41. "NASA's First Asteroid Deflection Mission Enters Next Design Phase," NASA.

42. Malik, "How Trump's Space Force Would Help Protect Earth From Future Asteroid Threats."

43. "NASA's First Asteroid Deflection Mission Enters Next Design Phase," NASA.

44. "NASA's First Asteroid Deflection Mission Enters Next Design Phase," NASA.

45. Michael Buckley, "Asteroid-Deflection Mission Passes Key Development Milestone," Phys.org, September 7, 2008, https://phys.org/news/2018-09-asteroid-deflection-mission-key-milestone.html.

46. Buckley, "Asteroid-Deflection Mission Passes Key Development Milestone."

47. Jeff Foust, "NASA Awards DART Launch Contract to SpaceX," SpaceNews, April 11, 2019, https://spacenews.com/nasa-awards-dart-launch-contract-to-spacex/.

48. Malik, "How Trump's Space Force Would Help Protect Earth From Future Asteroid Threats."

49. "National Near-Earth Object Preparedness Strategy and Action Plan," Interagency Working Group for Detecting and Mitigating the Impact of Earth-bound Near-Earth Objects of the National Science and Technology Council, June 2018, https://www.whitehouse.gov/wp-content/uploads/2018/06/National-Near-Earth-Object-Preparedness-Strategy-and-Action-Plan-23-pages-1MB.pdf, 6.

50. "National Near-Earth Object Preparedness Strategy and Action Plan," Interagency Working Group for Detecting and Mitigating the Impact of Earth-bound Near-Earth Objects of the National Science and Technology Council, 3–4.

51. "National Near-Earth Object Preparedness Strategy and Action Plan," Interagency Working Group for Detecting and Mitigating the Impact of Earth-bound Near-Earth Objects of the National Science and Technology Council, 3.

52. "National Near-Earth Object Preparedness Strategy and Action Plan," Interagency Working Group for Detecting and Mitigating the Impact of Earth-bound Near-Earth Objects of the National Science and Technology Council, 4.

53. "National Near-Earth Object Preparedness Strategy and Action Plan," Interagency Working Group for Detecting and Mitigating the Impact of Earth-bound Near-Earth Objects of the National Science and Technology Council, 5.

54. "National Near-Earth Object Preparedness Strategy and Action Plan," Interagency Working Group for Detecting and Mitigating the Impact of Earth-bound Near-Earth Objects of the National Science and Technology Council, 5.

55. "National Near-Earth Object Preparedness Strategy and Action Plan," Interagency Working Group for Detecting and Mitigating the Impact of Earth-bound Near-Earth Objects of the National Science and Technology Council, 6.

56. "National Near-Earth Object Preparedness Strategy and Action Plan," Interagency Working Group for Detecting and Mitigating the Impact of Earth-bound Near-Earth Objects of the National Science and Technology Council, 1.

57. "National Near-Earth Object Preparedness Strategy and Action Plan," Interagency Working Group for Detecting and Mitigating the Impact of Earth-bound Near-Earth Objects of the National Science and Technology Council, 8.

58. Amanda Jane Hughes, "Mining Asteroids Could Unlock Untold Wealth—Here's How to Get Started," The Conversation US, May 2, 2018, http://theconversation.com/mining-asteroids-could-unlock-untold-wealth-heres-how-to-get-started-95675.

59. "Water—The Key Resource in Space," Planetary Resources, accessed August 9, 2019, https://www.planetaryresources.com/products/.

60. Hughes, "Mining Asteroids Could Unlock Untold Wealth."

61. Leonard David, "Is Asteroid Mining Possible? Study Says Yes, for $2.6 Billion," Future US, Inc., April 24, 2012, https://www.space.com/15405-asteroid-mining-feasibility-study.html.

62. Whitwam, "SpaceX Wins Contract to Launch NASA's DART Asteroid Impactor."

63. Mack, "Trump's Space Force."

64. David, "Is Asteroid Mining Possible?"

65. Insinna, "Trump Officially Organizes the Space Force Under the Air Force...for Now."

66. Loren Grush, "NASA Is Opening the Space Station to Commercial Business and More Private Astronauts," Vox Media, Inc, June 7, 2019, https://www.theverge.com/2019/6/7/18656280/nasa-space-station-private-astronauts-commercial-business.

67. TIME, "President Trump Talks About Launching Space Force to Fight Wars in Space Alongside Military," YouTube, March 13, 2018, https://www.youtube.com/watch?time_continue=34&v=3T_nwIe8td4.
68. Mack, "Trump's Space Force."
69. Gabriella Muñoz, "Pence: Space Force Will Ensure U.S. Is 'Dominant' in Space as It Is on Earth," *Washington Times*, October 23, 2018, https://www.washingtontimes.com/news/2018/oct/23/mike-pence-space-force-war-fighting-domain/.
70. William J. Broad and David E. Sanger, "Flexing Muscle, China Destroys Satellite in Test," *New York Times*, January 19, 2007, https://www.nytimes.com/2007/01/19/world/asia/19china.html.
71. Carin Zissis, "China's Anti-Satellite Test," Council on Foreign Relations, February 22, 2007, https://www.cfr.org/backgrounder/chinas-anti-satellite-test.
72. Broad and Sanger, "Flexing Muscle, China Destroys Satellite in Test."
73. Zissis, "China's Anti-Satellite Test."
74. Mack, "Trump's Space Force."
75. Mack, "Trump's Space Force."
76. Jacqueline Klimas, "Neil deGrasse Tyson: Space Force Mission Should Include Asteroid Defense, Orbital Clean Up," Politico LLC, September 7, 2018, https://www.politico.com/story/2018/09/07/neil-degrasse-space-forceasteroid-defense-808976.
77. Neil deGrasse Tyson and Avis Lang, *Accessory to War: The Unspoken Alliance Between Astrophysics and the Military* (New York: W. W. Norton & Company, 2018), 254.
78. Klimas, "Neil deGrasse Tyson."
79. "Bucknam, Dr. Mark A. (Col USAF-Ret)," National War College, accessed August 12, 2019, https://nwc.ndu.edu/About/Faculty/ArticleView/Article/577420/bucknam-dr-mark-a-col-usaf-ret/.
80. Mark A. Bucknam, private interview with Defender Publishing agent, emailed document, May 29, 2019.
81. Mark Bucknam and Robert Gold, "Asteroid Threat? The Problem of Planetary Defence," *Survival: Global Politics and Strategy* 50, no. 5 (2008): 141–156.
82. Bucknam, private interview with Defender Publishing agent.
83. Bucknam and Gold, "Asteroid Threat? The Problem of Planetary Defence."
84. Bucknam, private interview with Defender Publishing agent.
85. Bucknam, private interview with Defender Publishing agent.
86. Bucknam, private interview with Defender Publishing agent.
87. Bucknam, private interview with Defender Publishing agent.

88. "The Great Midwest Wildfires of 1871," National Weather Service, accessed August 12, 2019, https://www.weather.gov/grb/peshtigofire2.

89. "Chicago Fire of 1871," A&E Television Networks, LLC, updated August 21, 2018, https://www.history.com/topics/19th-century/great-chicago-fire.

90. Dale Killingbeck, "Could a Meteorite or Comet Cause All the Fires of 1871?," meteorite-identification.com, August 23, 2004, http://meteorite-identification.com/mwnews/08232004.htm.

91. Irene Mona Klotz, "Did a Comet Trigger the Great Chicago Fire?," *Discovery News*, March 5, 2004, http://dsc.discovery.com/news/briefs/20040301/comet.html.

92. Klotz, "Did a Comet Trigger the Great Chicago Fire?"

93. Killingbeck, "Could a Meteorite or Comet Cause All the Fires of 1871?"

94. Klotz, "Did a Comet Trigger the Great Chicago Fire?"

95. Killingbeck, "Could a Meteorite or Comet Cause All the Fires of 1871?"

96. J. D. Thomas, "The Great Chicago Fire of 1871: Alternate Origin Stories," Accessible Archives Inc., October 8, 2013, https://www.accessible-archives.com/2013/10/chicago-fire-alternate-origin/.

97. Hannah Ellis-Petersen, "Philae Comet Lander Alien 'Cover-Up' Conspiracy Theories Emerge," *The Guardian*, November 13, 2014, https://www.theguardian.com/science/2014/nov/13/philae-comet-lander-alien-cover-up-conspiracy-theories-emerge.

98. Calla Cofield, "Why Does Comet 67P Sing? Scientists Think They Know," Space.com, August 21, 2015, https://www.space.com/30323-mystery-of-comet67p-singing-solved.html.

99. Ellis-Petersen, "Philae Comet Lander Alien 'Cover-Up' Conspiracy Theories Emerge."

100. Report sent to Tom Horn via email, Monday, December 22, 2014.

101. Jake Parks, "Organic Molecules Make Up Half of Comet 67P." *Astronomy*, December 1, 2017, http://www.astronomy.com/news/2017/12/comet-67p.

102. Tom Fish, "UFO Proof? Slowed-Down Comet P67 Audio Is 'ALIEN BROADCAST to Solar System,'" *The Express*, May 16, 2019, https://www.express.co.uk/news/weird/1124672/ufo-proof-comet-p67-audio-alien-broadcast-rosetta-ufo-sighting-scott-waring.

103. Ellis-Petersen, "Philae Comet Lander Alien 'Cover-Up' Conspiracy Theories Emerge."

104. Michael Salla, "Rosetta Comet Radio Signals and UFO Claims Spark Controversy," Exopolitics Institute News Service, September 29, 2014, https://exonews.org/rosetta-comet-radio-signals-ufo-claims-spark-controversy/.

105. Douglas Mangum, *Lexham Glossary of Theology* (Bellingham, WA: Lexham Press, 2014).
106. Donald W. Patten, *The Biblical Flood and the Ice Epoch* (Seattle: Pacific Meridian Publishing Company, 1966), chapter 8, https://www.creationism.org/patten/PattenBiblFlood/PattenBiblFlood08.htm.
107. Patten, *The Biblical Flood and the Ice Epoch*.
108. Flavius Josephus, *The Works of Flavius Josephus*, trans. William Whiston (London: Thomas Tegg, 1825), 19.

CHAPTER 4
PLANETARY PING-PONG

1. Carney, "Did a Comet Cause the Great Flood?"
2. Carney, "Did a Comet Cause the Great Flood?"
3. Carney, "Did a Comet Cause the Great Flood?"
4. Sandra Blakeslee, "Ancient Crash, Epic Wave," *New York Times*, November 14, 2006, https://www.nytimes.com/2006/11/14/science/14WAVE.html.
5. Carney, "Did a Comet Cause the Great Flood?"
6. Carney, "Did a Comet Cause the Great Flood?"
7. Blakeslee, "Ancient Crash, Epic Wave."
8. Carney, "Did a Comet Cause the Great Flood?"
9. Blakeslee, "Ancient Crash, Epic Wave."
10. "Diatom," *Encyclopædia Britannica*, accessed August 13, 2019, https://www.britannica.com/science/diatom.
11. Blakeslee, "Ancient Crash, Epic Wave."
12. Caitlin O'Kane, "Earth Is Approaching the Same 'Meteor Swarm' That May Have Caused an Entire Forest to Explode in 1908," CBS News, June 12, 2019, https://www.cbsnews.com/news/earth-is-approaching-taurid-meteor-swarm-tunguska-event-1908-caused-an-entire-forest-to-explode/.
13. Patten, *The Biblical Flood and the Ice Epoch*.
14. Patten, *The Biblical Flood and the Ice Epoch*, chapter 3.
15. Patten, *The Biblical Flood and the Ice Epoch*.
16. Patten, *The Biblical Flood and the Ice Epoch*, chapter 4.
17. Patten, *The Biblical Flood and the Ice Epoch*, chapter 6.
18. Patten, *The Biblical Flood and the Ice Epoch*, chapter 4.
19. Fraser Caine, "K-T Boundary," Universe Today, accessed September 8, 2019, https://www.universetoday.com/39801/k-t-boundary/.
20. *AF/A8XC Natural Impact Hazard (Asteroid Strike) Interagency Deliberate Planning Exercise After Action Report December 2008*, Directorate of Strategic Planning, Headquarters, United States Air

Force, accessed August 13, 2019, https://cneos.jpl.nasa.gov/doc/
Natural_Impact_After_Action_Report.pdf.

21. Thomas Horn and Stephen Quayle, *Unearthing the Lost World of
the Cloudeaters: Compelling Evidence of the Incursion of Giants,
Their extraordinary Technology, and Imminent Return* (Crane, MO:
Defender Publishing, 2017), 82.

22. Horn and Quayle, *Unearthing the Lost World of the Cloudeaters.*

23. Mindy Weisberger, "The Moon's Surface Is Totally Cracked,"
Live Science, April 23, 2019, https://www.livescience.com/65298-
impacts-cracked-the-moon.html.

24. Mike Wehner, "Massive Object Damaged Uranus Forever," *New
York Post*, July 3, 2018, https://nypost.com/2018/07/03/massive
-object-damaged-uranus-forever/.

25. "Bode's Law," *Encyclopædia Britannica*, accessed August 13, 2019,
https://www.britannica.com/science/Bodes-law.

26. Horn and Quayle, *Unearthing the Lost World of the Cloudeaters,*
98.

27. C. K. Quarterman, "Rahab the Home of Fallen Angels," End
Time Ministries, October 20, 2011, https://www.fallenangels-
ckquarterman.com/rahab/.

28. "Giant Planet Ejected From This Solar System," Southwest Research
Institute, November 10, 2011, https://www.swri.org/press-release/
giant-planet-ejected-solar-system.

29. Nola Taylor Redd, "The Late Heavy Bombardment: A Violent
Assault on Young Earth," Space.com, April 29, 2017, https://www.
space.com/36661-late-heavy-bombardment.html.

30. "P/Shoemaker-Levy 9," NASA, accessed August 13, 2019,
https://solarsystem.nasa.gov/asteroids-comets-and-meteors/comets/
p-shoemaker-levy-9/in-depth/.

31. Elizabeth Howell, "Shoemaker-Levy 9: Comet's Impact Left Its
Mark on Jupiter," Future US, Inc., January 24, 2018, https://www.
space.com/19855-shoemaker-levy-9.html.

32. Howell, "Shoemaker-Levy 9."

33. "P/Shoemaker-Levy 9," NASA.

34. Milton Kazmeyer, "Long Term Effects of an Asteroid Impact on
Earth," Hearst Seattle Media, LLC, accessed August 13, 2019,
https://education.seattlepi.com/long-term-effects-asteroid-impact
-earth-4601.html.

35. Kazmeyer, "Long Term Effects of an Asteroid Impact on Earth."

36. *AF/A8XC Natural Impact Hazard (Asteroid Strike) Interagency
Deliberate Planning Exercise After Action Report December 2008,*
Directorate of Strategic Planning, Headquarters, US Air Force.

37. "The Dangers of Crowds," Curiosity.com, May 13, 2016,
https://curiosity.com/topics/the-dangers-of-crowds-curiosity/.

38. "How the Hillsborough Disaster Unfolded," BBC, April 26, 2019, https://www.bbc.com/news/uk-19545126.
39. Mike Rothschild, "The Worst Human Stampedes," Ranker, accessed August 13, 2019, https://www.ranker.com/list/worst-human-stampedes-092415/mike-rothschild.
40. Julie Masis and Haroon Siddique, "Cambodia Water Festival Turns to Tragedy in Phnom Penh," *The Guardian*, November 22, 2019, https://www.theguardian.com/world/2010/nov/23/cambodia-water-festival-phnom-penh.
41. Rothschild, "The Worst Human Stampedes."
42. "'800 Dead' in Baghdad Bridge Disaster," *The Guardian*, August 31, 2005, https://www.theguardian.com/world/2005/aug/31/iraq.
43. Anjuli Sastry and Karen Grigsby Bates, "When LA Erupted in Anger: A Look Back at the Rodney King Riots," NPR, April 26, 2017, https://www.npr.org/2017/04/26/524744989/when-la-erupted-in-anger-a-look-back-at-the-rodney-king-riots.
44. Danieljbmitchel, "L.A. Riots of 1992: Rodney King Speaks; Late Troop Arrival," YouTube, September 29, 2007, https://www.youtube.com/watch?v=tgiR04ey7-M.
45. "Los Angeles Riots Fast Facts," Cable News Network, April 22, 2019, https://www.cnn.com/2013/09/18/us/los-angeles-riots-fast-facts/index.html.
46. *AF/A8XC Natural Impact Hazard (Asteroid Strike) Interagency Deliberate Planning Exercise After Action Report December 2008*, Directorate of Strategic Planning, Headquarters, US Air Force.
47. John Orrell, "Hurricane Katrina Response: National Guard's 'Finest Hour,'" US Army, August 27, 2010, https://www.army.mil/article/44368/hurricane_katrina_response_national_guards_finest_hour.
48. "FEMA Faces Intense Scrutiny," NewsHour Productions LLC, September 9, 2005, https://www.pbs.org/newshour/politics/government_programs-july-dec05-fema_09-09.
49. William H. Thiesen, "The Long Blue Line: Coast Guard Responders During Record-Setting Hurricane Katrina," Coast Guard Compass Archive, August 25, 2016, https://coastguard.dodlive.mil/2016/08/the-long-blue-line-coast-guard-responders-during-record-setting-hurricane-katrina/.
50. Orrell, "Hurricane Katrina Response."
51. "FEMA Faces Intense Scrutiny," PBS.org.
52. "14 Days: A Timeline," FRONTLINE, accessed August 13, 2019, https://www.pbs.org/wgbh/pages/frontline/storm/etc/cron.html.
53. Dina Spector, "Here's Why Astronomers Did Not Detect the Russia Meteor Ahead of Time," *Business Insider*, February 15, 2013, https://www.businessinsider.com/

heres-why-astronomers-did-not-detect-the-russia-meteor-ahead-of-time-2013-2.

54. Eric Mack, "NASA Head: Expect a Major Asteroid Strike in Your Lifetime," CNET, April 30, 2019, https://www.cnet.com/news/nasa-head-expect-a-major-asteroid-strike-in-your-lifetime/.

55. Spector, "Here's Why Astronomers Did Not Detect the Russia Meteor Ahead of Time."

56. Mack, "NASA Head."

CHAPTER 5
HIS NAME IS WORMWOOD

1. J. F. Walvoord and R. B. Zuck eds., *The Bible Knowledge Commentary: An Exposition of the Scriptures by Dallas Seminary Faculty, New Testament Edition* (Wheaton, IL: Victor Books, 1983), 953.

2. Though I stray often from Michael Heiser's SitchinIsWrong.com website to represent my opinion on Zecharia Sitchin's flawed theories, many thanks are due to Heiser for his concise list and well-documented presentation on what errors Sitchin made in his work.

3. Zecharia Sitchin, "The Official Website of Zecharia Sitchin," accessed August 13, 2019, http://www.sitchin.com/.

4. Michael Heiser, SitchinIsWrong, accessed August 13, 2019, http://www.sitchiniswrong.com/mshcv.htm.

5. Michael Heiser, "The Myth of a Sumerian 12th Planet: 'Nibiru' According to the Cuneiform Sources," SitchinIsWrong, accessed August 13, 2019, http://www.sitchiniswrong.com/nibirunew.pdf.

6. Heiser, "The Myth of a Sumerian 12th Planet."

7. Heiser, "The Myth of a Sumerian 12th Planet."

8. Heiser, "The Myth of a Sumerian 12th Planet."

9. FringePop321, "Wormwood and Planet X (Nibiru)—Is This the End?" YouTube, December 10, 2018, https://www.youtube.com/watch?v=We__lU1KFd8.

10. G. K. Beale, *The Book of Revelation: The New International Greek Testament Commentary* (Grand Rapids, MI; W.B. Eerdmans; Carlisle, Cumbria: Paternoster Press, 1999), 218.

11. Bible Study Tools, s.v. *"eisi,"* accessed August 13, 2019, https://www.biblestudytools.com/lexicons/greek/nas/eisi.html.

12. Geoffrey W. Bromiley, *Theological Dictionary of the New Testament,* eds. Gerhard Kittel and Gerhard Friedrich (Grand Rapids, MI: William B. Eerdmans Publishing Company, 1985), 86.

13. G. R. Beasley-Murray, "Revelation," *The New Bible Commentary*, eds. Gordon J. Wenham et al. (Downers Grove, IL: IVP Academic, 1994), 1,438.
14. Beale, *The Book of Revelation*, 479.
15. FringePop321, "Wormwood and Planet X (Nibiru)," textual emphasis was added in the moments Heiser applied vocal emphasis.
16. Beale, *The Book of Revelation*, 475.
17. Beale, *The Book of Revelation*, 475.
18. Beale, *The Book of Revelation*, 476; emphasis added.
19. Beale, *The Book of Revelation*, 478.
20. Beale, *The Book of Revelation*, 479.
21. Beale, *The Book of Revelation*, 480.

CHAPTER 6
AS IT WAS IN THE DAYS OF PHARAOH . . . ETYMOLOGY OF "WORMWOOD"

1. Horn and Quayle, *Unearthing the Lost World of the Cloudeaters*, Kindle locations 7082–7127.
2. David Noel Freedman ed., "Wormwood," *The Anchor Yale Bible Dictionary*, vol. 6 (New York: Doubleday, 1992), 973.
3. Freedman, "Wormwood."
4. Chad Brand et al., eds., "Wormwood," *Holman Illustrated Bible Dictionary* (Nashville: Holman Bible Publishers, 2003) 1686.
5. J. A. Strong, *A Concise Dictionary of the Words in the Greek Testament and the Hebrew Bible*, vol. 2 (Bellingham, WA: Logos Bible Software, 2009), 60.
6. John Henry, "The Chronological Order of The Revelation of Jesus Christ," prophecy.landmarkbiblebaptist.net, August 20, 2019, http://prophecy.landmarkbiblebaptist.net/rev-chronology.html.
7. R. A. Coombes, "Why the Book of Revelation Is Not Completely in Chronological Order," Alpha Omega Report, May 29, 2009, http://www.alphaomegareport.com/why-the-book-of-revelation-is-not-completely-in-chronological-order-52909/.
8. Coombes, "Why the Book of Revelation Is Not Completely in Chronological Order."
9. Coombes, "Why the Book of Revelation Is Not Completely in Chronological Order."
10. Louis Kondor ed., *Fatima in Lucia's Own Words: Sister Lucia's Memoirs*, trans. Dominican Nuns of Perpetual Rosary (n.p.: Fundação Francisco e Jacinta Marto, 2016), 123–124.
11. Sister Maria Lúcia, in a personal letter dated November 8, 1989, as quoted by "Congregation for the Doctrine of the Faith: The Message of Fátima," Vatican, accessed August 14, 2019, http://www.

vatican.va/roman_curia/congregations/cfaith/documents/rc_con_
cfaith_doc_20000626_message-fatima_en.html.

12. Dick Ahlstrom, "Chernobyl Anniversary: The Disputed Casualty
    Figures," *Irish Times*, April 2, 2016, https://www.irishtimes.com
    /news/world/europe/chernobyl-anniversary-the-disputed-casualty
    -figures-1.2595302.

13. *The Exodus Decoded*, directed by James Cameron, created by
    Simcha Jacobovici, aired August 20, 2016, on the History channel.

14. *The Exodus Decoded*, 27:32–40:23.

15. *The Exodus Decoded*, 40:32–40:39.

16. *The Exodus Decoded*, 40:29–41:11.

17. *The Exodus Decoded*, 41:26–42:51.

18. *The Exodus Decoded*, 45:08–47:20.

19. *The Exodus Decoded*, 48:14–48:45.

20. Joseph S. Byrnes and Leif Karlstrom, "Anomalous K-Pg–
    aged Seafloor Attributed to Impact-Induced Mid-Ocean
    Ridge Magmatism," *Science Advances* 4, no. 2 (February
    7, 2018), https://advances.sciencemag.org/content/4/2/
    eaao2994?utm_source=VancePak+%28updated+6%2F
    30%2F2017%29&utm_campaign=caf5d5a016-EMAIL_
    CAMPAIGN_2018_02_02&utm_medium=email&utm
    _term=0_56c46682ac-caf5d5a016-126548837.

21. Julia Rosen, "Large Impacts May Cause Volcanic Eruptions,"
    60-Second Science (podcast), Scientific American, May 16, 2017,
    https://www.scientificamerican.com/podcast/episode/large-
    impacts-may-cause-volcanic-eruptions/.

22. Roff Smith "Here's What Happened the Day the Dinosaurs
    Died," *National Geographic*, June 11, 2016, https://www.
    nationalgeographic.com/news/2016/06/what-happened-day-
    dinosaurs-died-chicxulub-drilling-asteroid-science/.

23. As quoted in Alwyn Scarth, *Vulcan's Fury: Man Against the
    Volcano* (New Haven, CT: Yale University Press, 1999), 248–249.

24. "1986: Hundreds Gassed in Cameroon Lake Disaster," BBC, August
    21, 1986, http://news.bbc.co.uk/onthisday/hi/dates/stories/august/21
    /newsid_3380000/3380803.stm.

25. Peter J. Baxter, M. Kapila, and D. Mfonfu, "Lake Nyos Disaster,
    Cameroon, 1986: The Medical Effects of Large Scale Emission of
    Carbon Dioxide?," *British Medical Journal* 298 (May 27, 1989):
    1437–1441.

26. Joshua Bearman and Allison Keeley, "The Mad Scramble to Claim
    the World's Most Coveted Meteorite," *Condé Nast*, December 17,
    2018, https://www.wired.com/story/scramble-claim-worlds-most-
    coveted-meteorite/.

27. "Peru: Doctors Aid in Rising Number of Illnesses after Meteorite Crash," Living in Peru, September 19, 2007, https://www. livinginperu.com/news-4732-environmentnature-peru-doctors-aid-in-rising-number-of-illnesses-after-meteorite-crash/.

28. Bearman and Keeley, "The Mad Scramble to Claim the World's Most Coveted Meteorite."

29. "Peru Meteor Illness Deepens," *The Age*, September 19, 2007, https://www.theage.com.au/world/peru-meteor-illness-deepens -20070919-ge5unv.html.

30. John D. Barry et al., ed., *The Lexham Bible Dictionary* (Bellingham, WA: Lexham Press, 2016).

31. Philo, *De Vita Mosis*, cf. 1.100.

32. This estimation is popular among countless world-population calculation sources online at the time of this writing. As merely one example: Brian Wang, "World Population Will Pass 8 Billion in 2023," Next Big Future, September 30, 2018, https:// www.nextbigfuture.com/2018/09/world-population-will-pass-8- billion-in-2023.html.

33. Walvoord and Zuck, *The Bible Knowledge Commentary*, 953.

34. Blue Letter Bible, s.v. "*theiōdēs*," accessed August 14, 2019, https:// www.blueletterbible.org/lang/lexicon/lexicon .cfm?Strongs=G2306&t=KJV.

35. Blue Letter Bible, s.v. "*hyakinthinos*," accessed August 14, 2019, https://www.blueletterbible.org/lang/lexicon /lexicon.cfm?Strongs=G5191&t=KJV.

INDEX